MAINE'S FAVORITE BIRDS

JEFFREY V. WELLS & ALLISON CHILDS WELLS

ILLUSTRATED BY EVAN BARBOUR

Down East Books
Camden, Maine

Down East Books

An imprint of Globe Pequot

Distributed by NATIONAL BOOK NETWORK

First paperback edition: June 2012
First Down East edition: June 2023

Library of Congress Cataloging-in-Publication Data

Wells, Jeffrey V. (Jeffrey Vance), 1964-
  Maine's favorite birds / Jeffrey V. Wells and Allison Childs Wells ; illustrated by Evan W. Barbour. — 1st pbk. ed.
     p. cm.
  Includes index.
  ISBN 978-1-68475-211-9 (pbk. : alk. paper)
  1. Birds—Maine—Identification. 2. Bird watching—Maine—Guidebooks. I. Wells, Allison Childs. II. Barbour, Evan W III. Title.
  QL684.M2W45 2012
  598.072'34741—dc23
                      2012007638

Cover art (Baltimore Orioles, Yellow Warbler, Barn Swallow, Black-capped Chickadee, American Robin, Atlantic Puffin, and Common Loon) by Evan Barbour
Designed by Geraldine Millham, Westport, Massachusetts
Copyedited by Genie Dailey, Fine Points Editorial Services, Jefferson, Maine

∞™ The paper used in this publication meets the minimum requirements of American National Standard for Information Sciences-Permanence of Paper for Printed Library Materials, ANSI/NISO Z39.48-l 992.

FOR OUR SON

ACKNOWLEDGMENTS

With love and gratitude, we thank our maternal grandmothers, Audrey Giles Chase and Ida Moore Cuthbertson, for sparking our interest in birds at an early age, and our parents, Arthur Wells and Konni Chase Wells and Dana and Jewell Childs, for nurturing this interest as we were growing up. Special thanks also to many people who had a very positive impact early in our birding lives: Jim Miller; Inez Boyd, Marjorie Cookson, Lorna Evans, Gail Freese, and other members of the Bangor Nature Club and the Penobscot Valley Audubon Society during the 1970s and '80s.

We also wish to express our appreciation to the early pioneers of bird identification and conservation, especially Roger Tory Peterson and John James Audubon, whose works made birds, birding, and bird conservation accessible to the general public. Their contributions will forever live on in the many field guides and other bird identification tools in use today and in those yet to come.

CONTENTS

INTRODUCTION

Maine is a spectacularly beautiful state, a place of lush forests, rocky coastlines, and sandy beaches. Lakes, rivers, and streams in all shapes and sizes reach across the landscape. Mountains and rolling fields abound. Such an extensive range of natural beauty makes Maine a magnet for nature lovers far and wide.

The diverse habitats also make Maine a hot spot teeming with opportunities to see and hear some of North America's most iconic birds—Bald Eagles soaring over a frozen river lined with ice-fishing shacks; Atlantic Puffins peppering an offshore island draped in summer sun; Common Loons, their primeval cries echoing across a quiet lake. These sights and sounds are integral to Maine's identity and essential to the reason why millions of tourists visit the state in a typical year. They also contribute greatly to the quality of life for those of us fortunate to call Maine our home.

As native Mainers whose ancestors arrived here centuries ago, our appreciation for the nature of our state runs deep and strong. Like millions of others, we take a special delight in birds. Their beautiful plumages, captivating songs, and fascinating behaviors make them relatively easy to find, watch, and enjoy.

There are now dozens of excellent field guides to North American birds that provide details of more than 800 species that have been recorded in the United States and Canada. At the same time, the wide scope of such references can be overwhelming to the beginning birder or mild enthusiast who simply wants to identify the birds they most often see or are most likely to find.

In this book, we celebrate Maine's favorite birds. With more than 400 bird species documented as occurring in Maine, keeping our list manageable was no easy task, especially when no two lists of "favorites" may be exactly the same. Few could resist including colorful beauties like Northern Cardinal and Eastern Bluebird. Some birders, though, might include range-restricted "Maine specialties" such as the mountaintop-dwelling Bicknell's Thrush or the famously elusive Spruce Grouse, a bird of the North Woods. Although tracking down hard-to-find species like these yields great rewards, doing so requires considerable effort and a high level of expertise—two important reasons why the majority of Maine residents and visitors who enjoy birds do not pursue such species, and why we did not include them in this book. Likewise, although we reference in the Birding in Maine section birds for which a particular birding hot spot is known, those that did not meet our objectives for this book were not included in the species accounts.

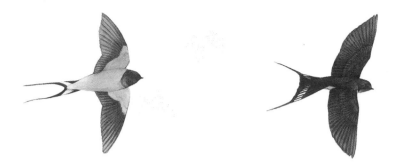

For the majority of people, most of their birding pleasure comes from watching birds in their backyards, gardens, parks, local ponds, and other areas they visit frequently. They love "their" chickadees, loons, and herons. They want to know how to identify the different sparrow species they see at their feeders. They enjoy learning that a Blue Jay is not a bluebird—even though it is a blue bird. They want to be able to help their children tell one gull species from another during a family outing to the coast, and to help them discover a bit more about the many species that visit the family bird feeder.

Those are the kinds of experiences we hope to enhance by creating this easy-to-use guide. Our goal was also to produce a book that is beautiful, something resident bird enthusiasts will use often and visitors to Maine will treasure as a keepsake. For those reasons, we hope you enjoy the elegant illustrations of bird artist Evan Barbour as much as we do.

Ornithologists and skilled birders may note that the birds are not presented in the book in strict official taxonomic order. The order and groupings of species is instead based on ways we thought would make the most sense for beginners to find species that either would occur in similar habitats or that look similar.

To make identification as easy as possible, we have included additional materials, including tips for identifying birds and some of the wonderful places to find them. We also provided information about some of the ways adults and children alike can take action for birds and even put their bird sightings to use for science and conservation.

Above all else, in *Maine's Favorite Birds* we hope you discover a book that will help you see birds—and Maine—in a new, more fascinating way.

Good birding,
Allison and Jeff Wells

TOOLS OF THE BIRDING TRADE

One of the best things about birding is that you can bird just about any way you wish. If you keep your eyes and ears ready, you will very likely see birds, but having a few key pieces of equipment on hand can make a big difference in your enjoyment of birds.

Most important among your tools is a good field guide. To people who love nature and discovery, few things are as satisfying as identifying the creature they are observing. *Maine's Favorite Birds* will help you quickly and easily identify birds that are familiar to most people here in Maine. As you gain advanced birding skills, there are many field guides out there that can assist you. The more serious you get, the more references you may wish to add to your own personal library.

Using a pair of binoculars will allow you to see and identify more birds from a distance. Having them within easy reach will also open a whole new world of detail that you may not have known exists. You may have grown up seeing Blue Jays, for example, and now rarely give them a second glance. But seen through a pair of binoculars, those black and white markings against the blue feathers are so stunning that you may never take them for granted again.

We also recommend adding to your list of essential equipment a digital camera, even if it is an inexpensive model, since even a not-so-great photo of a bird will allow you to identify it later with the help of reference guides. Most cameras now record audio and video, which means you can get a recording of a bird's vocalization, making the species more easily identifiable.

Once you are hooked on birding, you may want to add a birding telescope to your equipment. Read reviews of the different models, and, if possible, personally try out any you may be interested in before purchasing one. Many people now carry on their iPhones one of various bird identification apps. There are countless CDs and websites at your fingertips that help you identify birds by their sounds; many can be uploaded to an iPod for reference in the field.

If you ask any experienced birder how best to improve your birding skills, he or she will tell you, "Time in the field." The more you get out and bird, the more you will see and the more you will learn about the subtleties of habitat types that different species prefer, which birds stay in Maine year-round, which leave for the winter, when and where certain species are most abundant, and much more. If you can tag along with someone who is more advanced than you are at bird identification and who likes to share his or her knowledge with others, you will learn even more quickly. Good birders willing to share their knowledge are the best "tools" of all.

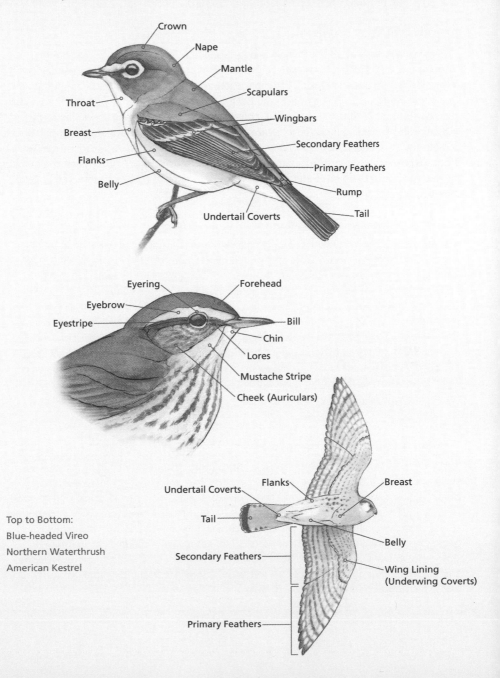

LEARNING THE FIELD MARKS

Many of the field marks described in the species accounts in this book refer to specific anatomical parts of a bird. To understand those field marks, here is a "map" showing the parts of a bird that are most important for understanding field marks. Getting to know the anatomy of a bird will help you identify birds more quickly and accurately.

Crown

Nape

Mantle

Scapulars

Throat

Wingbars

Breast

Secondary Feathers

Flanks

Primary Feathers

Belly

Rump

Undertail Coverts

Tail

Eyering

Forehead

Eyebrow

Eyestripe

Bill

Chin

Lores

Mustache Stripe

Cheek (Auriculars)

Flanks

Breast

Undertail Coverts

Tail

Belly

Secondary Feathers

Wing Lining
(Underwing Coverts)

Primary Feathers

Top to Bottom:
Blue-headed Vireo
Northern Waterthrush
American Kestrel

ATLANTIC PUFFIN *(Fratercula arctica)*

One of Maine's most sought-after species, the Atlantic Puffin is often called "parrot of the sea." In the summer breeding season, bill is a showy protrusion of red, blue, and yellow contrasting strongly with white face, black cap and upperparts. Undersides white. Legs bright orange. Best seen from one of many boats giving tours to offshore breeding islands, especially Eastern Egg Rock in Muscongus Bay, where reintroduction efforts starting in 1973 have reestablished a thriving breeding colony. **Voice**: Buzzy, chainsaw-like growl made primarily in breeding burrow **Length**: 11.5–13.5" (29–34 cm)

DOUBLE-CRESTED CORMORANT *(Phalacrocorax auritus)*

A common sight along Maine's coast and larger inland lakes and rivers, from spring through late fall. A large, black waterbird with a long, thin neck. Orange throat and bill. When swimming, especially at a distance, often mistaken for Common Loon, but note that cormorants tilt the bill upwards and have a different shape to the bill and head. When seen close-up, note the almost startling emerald-green eye. Adults in spring breeding plumage show two small crests, a feature rarely seen. Immature birds show whitish breast but otherwise are dark brownish. In winter, most Double-crested Cormorants migrate from Maine to the southern U.S., replaced in Maine by the larger Great Cormorant, which shows a yellow throat pouch and white flank patches. **Voice**: Hoarse grunts, rarely heard. **Length**: 28–35" (70–90 cm)

HERRING GULL *(Larus argentatus)*

The Herring Gull is the bird commonly referred to as "seagull" and is essential to any classic coastal Maine scene. Bold and brazen, they are known to help themselves to the contents of picnic baskets and food left unattended at tables. White undersides, white head (streaked with brown in winter), pale gray back and upper wings with black wing-tips. Pink legs. Yellow bill has red spot near tip of lower mandible. Takes four years to reach adulthood. Immature plumages are mottled brown with dark bill, legs, and eyes. Coloration lightens each year toward maturity. The larger Great Black-backed Gull has a black back and wings in adult plumage. The smaller Laughing Gull has a black head and reddish bill in adult breeding plumage. Species most easy to confuse with the Herring Gull is the Ring-billed Gull, but note its smaller size, yellow legs, and black ring on bill in adult plumage. **Voice**: Loud, trumpeting call—the iconic sound of the "seagull." **Length**: 22–26" (56–66 cm)

COMMON EIDER *(Somateria mollissima)*

Very common sea duck found along the Maine coast, in calm coves and along wild, rocky shorelines alike. Large and stocky with distinctive long, sloping bill. Male in breeding plumage has black belly, wings, and tail contrasting with snowy white back, breast, and head, set off with a black cap. Green tones on back of head. Female is brown barred with black. Immature male brownish-black with some white at neck, breast, and sometimes back. Usually seen in flocks (called "rafts"), sometimes with hundreds of individuals, especially in late fall and winter. In early summer, tiny ducklings can sometimes be seen among the females that are watching over them. **Voice**: Hoarse cooing and grunting, rarely heard. **Length**: 19–28" (50–71 cm)

ATLANTIC PUFFIN

DOUBLE-CRESTED CORMORANT

Great Black-backed Gull

Ring-billed Gull

HERRING GULL

immature

adult

Laughing Gull

♂

COMMON EIDER

♀

CANADA GOOSE *(Branta canadensis)*

Large waterbird, common and familiar even to beginning birders. Brownish back; long black neck and head with white chin. In spring, Canada Geese may be seen in fields, on lawns of parks, and on other grassy areas, often with chicks tagging along. Formerly seen mostly during spring and fall migration, non-migratory birds have been introduced and are common nesters and year-round residents wherever they can find open water and sufficient food. **Voice**: A loud *honk.* **Length**: 30–43" (76–110 cm)

MALLARD *(Anas platyrhynchos)*

Perhaps the most familiar of ducks found in Maine, seen in ponds, lakes, rivers, swamps, and coastal bays and marshes. A large dabbler, male is easily recognized by metallic green head and neck, reddish breast, and white "necklace." Females overall are pale brown with white-bordered speculum and orange bill. In summer, male looks similar to female but with rusty wash on breast and yellow bill; both easily confused with American Black Duck (see below). **Voice**: The classic duck *quack.* **Length**: 23" (58 cm)

AMERICAN BLACK DUCK *(Anas rubripes)*

Large, blackish-brown duck. Face and foreneck paler than body. Purple speculum on wing. Male has yellow bill; female's is olive. Compare to similar-looking female Mallard, which shows white speculum border, orange bill, and is paler overall; summer-plumaged male Mallard has rusty wash on breast, white edge to speculum, and also is paler overall. Interbreeds with Mallards; hybrids not uncommon. Like Mallard, can occur in a wide variety of wetlands and water bodies, but usually more reclusive. In early winter, large numbers move south from Canada to Maine's coast for the winter or for stopover until cold and ice conditions force them to move farther south. **Voice**: A raspy *quack,* very similar to but slightly lower pitched than Mallard's. **Length**: 23" (58 cm)

WOOD DUCK *(Aix sponsa)*

One of Maine's most strikingly beautiful and distinctive duck species. Males have a long, iridescent green crest set off by white stripes, and a white collar and chin-strap. Bill is reddish-orange with a dark tip. The deep reddish chest is separated from the yellowish sides by a bold black stripe. Back is black with metallic sheen. Females are dull gray-brown with smaller crest and white around eye. Prefers edges of wooded swamps and marshes. Nests in holes in trees and will also use large nest boxes. Usually returns from wintering grounds in March. **Voice**: A loud, high-pitched whistled *wheep.* **Length**: 18.5" (47 cm)

HOODED MERGANSER *(Lophodytes cucullatus)*

Small black-and-white diving duck with thin bill for catching small fish, crayfish, and other small aquatic invertebrates. Often occurs in still waters of streams, lakes, ponds, and coastal bays. Male shows black-and-white pattern but with rusty sides. When displaying, opens a distinctive white fan-shaped crest. When not displaying, the crest is lowered and white is reduced to a small line. Females are brownish overall with shorter crest and a dull, yellowish bill. Like the Wood Duck, this species nests in holes in trees and will also use large nest boxes. Unlike the Wood Duck, this species regularly winters in small flocks in southern coastal Maine. **Voice**: Variety of low croaking sounds but infrequently heard. **Length**: 18" (46 cm)

Mallard

American
Black Duck

speculum

♀

♂

CANADA GOOSE

MALLARD

AMERICAN BLACK DUCK

♀

♂

WOOD DUCK

♂

HOODED MERGANSER

♀

GREEN HERON *(Butorides virescens)*

Small, stocky, short-legged summer-resident heron. Greenish-blue upperparts; chestnut on sides of neck. Legs are bright orange-yellow in breeding males. Juveniles are brownish above, heavily streaked below. Often seen perched on snags above ponds or flying across the water. **Voice**: A loud, piercing *keeyow.* **Length**: 18" (46 cm) **Wingspan**: 25" (66 cm)

GREAT BLUE HERON *(Ardea herodias)*

The most commonly seen large heron, sometimes called "crane" by those who may not have ever seen the even larger and uncommon-in-Maine Sandhill Crane. Large, grayish-blue wading bird. Black stripe above the eye, white foreneck with black streaks. Yellow bill and showy plumes during breeding plumage; nonbreeders and juveniles lack plumes. Juvenile has dark crown. In flight, legs are extended and neck is tucked; wings make a heavy sweeping motion. Look for them wading at the marshy edges of ponds and lakes; along the edges of coastal bays, coves, and salt marshes; and also in streams and along river edges. A few linger in south-coastal Maine in winter but most migrate further south. **Voice**: Loud, hoarse *croak.* **Length**: 46" (117 cm) **Wingspan**: 70" (178 cm)

WILD TURKEY *(Maleagris gallopavo)*

Unmistakable—no other Maine species compares in size and shape. Males have a reddish head; females and immature, a brownish head. Apparently extinct in Maine (and much of its North American range) by the early 1800s because of loss of habitat and overhunting, Maine's Inland Fisheries and Wildlife Agency began a reintroduction program in the early 1970s, introducing 41 Wild Turkeys from Vermont, releasing them in southern Maine. The program has been so wildly successful that it is not uncommon to see flocks of dozens of the birds feeding in backyards and fields, and there are now spring and fall hunting seasons. **Voice**: Familiar gobble calls heard especially in spring during mating season **Length**: Female 37" (94 cm), male 46" (117 cm)

GREEN HERON

juvenile

adult

GREAT BLUE HERON

adult

juvenile

WILD TURKEY

BALD EAGLE *(Haliaeetus leucocephalus)*

Known and loved as the emblem of the United States, the Bald Eagle is also a symbol of conservation success. The species returned from perilously low numbers in the lower forty-eight states, including Maine, thanks to strong management efforts and banning of the pesticide DDT. Adults show classic white head and tail offset by massive brown body. Pairs reuse their huge nests each year, typically on coastal islands or near large water bodies. Immatures, dark brown with splotches of white in underwings, take five years to reach adulthood. In Maine, they winter along coast and near large rivers, often perched above frozen water near ice-fishing shacks in search of discarded fish remains. In spring, a dozen or more Bald Eagles may congregate with Osprey and gulls to feast on spring runs of alewives migrating from the ocean to spawn in inland lakes. **Voice**: Shrill, high-pitched piping. **Length**: 28–37" (71–96 cm) **Wingspan**: 72–90" (183–244 cm)

COMMON MERGANSER *(Mergus merganser)*

A rather large and long-bodied duck with a long, narrow, saw-toothed bill adapted for catching fish (a characteristic common to all mergansers). Males seen on the water show bright, crisp white bodies and black backs with dark green heads and red bill. Females and immatures are gray bodied with brown, shaggy heads and red bills. Common Mergansers are regularly seen on inland lakes and rivers, even feeding in small areas of open water in otherwise completely frozen rivers in the dead of winter. In early spring birds congregate in open stretches of larger rivers, sometimes numbering in the thousands, before migrating farther north. Nests in cavities in trees near lakes and large rivers where, in early summer, it is not uncommon to see a female with her downy young. **Voice**: Low grunts and croaks, rarely heard. **Length**: 25" (64 cm)

COMMON LOON *(Gavia immer)* ▾

A quintessential bird of Maine lakes, with a melancholy wail that rarely escapes comment by those who hear it. Breeding plumage shows black head and neckband contrasting with bold, black-and-white checkered pattern on back. Strong, dark, pointed bill. In Maine, common winterer along coast, when plumage lacks black and is gray above and light below and bill becomes lighter colored. Susceptible to poisoning caused by ingestion of lead fishing tackle and to mercury poisoning tied to air emissions from coal-fired power plants in the Midwest that drift eastward to Maine. Young sometimes seen riding on back of parent. **Voice**: Loud, "laughing" yodels and melancholy wails. **Length**: 26–35" (66–91 cm)

PIED-BILLED GREBE *(Podilymbus podiceps)*

Small grayish-brown waterbird with a fluffy white rear. Black forehead and throat. Light-colored bill is rather stubby and rounded with a black ring in breeding season. Found throughout Maine in marshy areas of ponds and lakes, more widely during migration, in scattered locations during the breeding season. **Voice**: During the breeding season gives a loud, maniacal, yodeled, *CAoo-CAoo-CAoo-CAoo,* dropping in pitch on the second syllable. Even when calling, they can be hard to see. **Length**: 13.5" (34 cm)

adult

immature

BALD EAGLE

COMMON MERGANSER

♀

♂

COMMON LOON

PIED-BILLED GREBE

TURKEY VULTURE *(Cathartes aura)*

Large, black bird most often seen soaring high in the sky. Silver flight feathers contrast with black wing lining, a field mark easily noted when bird is in flight. Red, mostly unfeathered head noticeable only at close range. Soaring Turkey Vultures hold wings upward in a shallow V shape, compared to Bald Eagles, which hold their wings flat in soaring flight. Until thirty or forty years ago, Turkey Vultures were rare in Maine and virtually unknown to most Mainers. Now commonly seen throughout most of the state except during dead of winter, though in recent years a few have remained and survived. Turkey Vultures begin migrating back into the state from farther south as early as February, often in flocks that are known to birders as "kettles." **Voice**: Essentially silent but can produce hissing sounds and clicks. **Length**: 27" (69 cm) **Wingspan**: 66–70" (170–178 cm)

OSPREY *(Pandion haliaetus)*

One of the most beloved and watched-over of our Maine birds, in part because they build their large stick nests at the tops of trees or poles where they are easily observed. A large fish-eating hawk with white belly and breast contrasting against dark upper wings and back. White head with dark brown stripe extending back from bold, yellow eye. When soaring, wings are bent into the shape of an M. Underwing shows bold dark patch at bend of wing. Ospreys begin returning to Maine in April from wintering grounds as far south as northern South America, one of the most anticipated birds of spring. Frequently seen carrying fish back to the nest in talons. **Voice**: A loud, persistent, and high-whistled *teeyerp,* often given in a series—one of the common summer sounds of Maine's coasts, lakes, and rivers. **Length**: 23–25" (56-64 cm). **Wingspan**: 59–70" (150–180 cm)

RED-TAILED HAWK *(Buteo jamaicensis)*

As its name suggests, this large hawk's best identification feature is its broad but short red tail. When perched, note its brown back and white undersides with darker band across belly. In flight, undersides of wings largely white with dark wingtips and thin, dark trailing edge and dark shoulder bar. Immatures are more streaked below and have brown tail with dark bands. A common hawk in Maine, often seen sitting on telephone wires and trees along highways and soaring over open fields. **Voice**: A raspy screamed *kree-aa.* **Length**: 19" (56 cm) **Wingspan**: 44–52" (114–132 cm)

TURKEY VULTURE

head-on view

OSPREY

RED-TAILED HAWK

adult

juvenile

SHARP-SHINNED HAWK *(Accipiter straitus)*

Small hawk, about the size of a Blue Jay or American Robin (but longer-tailed). Adult shows blue-gray back, wings, tail, neck, and crown, and light-colored undersides with fine, reddish barring except under tail. In flight, the short, rounded wings and thin, square-ended tail are distinctive, as is its habit of quickly flapping wings then gliding. As in all hawks and owls, females are distinctly larger than males. In immature birds, the blue-gray colors are replaced with brown. Similar Cooper's Hawk has proportionately longer, more rounded tail and larger head. Both Sharp-shinned and Cooper's hawks survive by catching small birds for food and are sometimes seen swooping by bird feeders in the hopes of startling prey. During fall migration, it is not uncommon to see dozens, even hundreds, of Sharp-shinned Hawks flying southward past hawk migration lookouts. **Voice**: A series of short, piercing *kik-kik-kik-kik* notes generally only heard near nest. **Length**: 10.5" (27 cm) **Wingspan**: 16" (43 cm)

COOPER'S HAWK *(Accipiter cooperii)*

Like a larger version of the Sharp-shinned Hawk, but with a proportionately longer, rounded (rather than square-tipped) tail. Cooper's Hawks also have a larger head and typically look like they have a dark cap because the back of the neck is lighter than in Sharp-shinned Hawk. This is one of the most common year-round resident hawks throughout the southern half of the state. **Voice**: Alarm call heard near nest is a series of short, sharp *kek* notes. **Length**: 14–20" (36–51 cm) **Wingspan**: 24–35" (62–90 cm)

AMERICAN KESTREL *(Falco sparverius)*

A small hawk similar in size to the Sharp-shinned Hawk. In contrast to Sharp-shinned and Cooper's hawks, American Kestrels in flight show pointed rather than rounded wings—a characteristic common to all falcons. Reddish back and tail as well as dark "teardrop" stripes on face are distinctive. Males have light blue on wings, replaced with brown in females and immatures. Open-country birds, often seen near farm fields and along roads and highways where they watch from telephone wires for their prey of small mammals and insects. When actively hunting, often hover above their prey before plunging feet first for the capture. Originally nested only in tree cavities, but now also nest in special nest boxes put up for them by bird lovers; sometimes nest in abandoned barns or other buildings. **Voice**: A series of rapid *klee-klee-klee* notes, given when excited; usually heard only near nesting sites. **Length**: 10.5" (27 cm) **Wingspan**: 20–25" (51–63 cm)

SHARP-SHINNED HAWK

juvenile ♀

juvenile ♀

adult ♂

adult ♂

COOPER'S HAWK

juvenile
♀

juvenile ♀

adult ♂

adult ♂

♀

♂

♂

AMERICAN KESTREL

KILLDEER *(Charadrius vociferus)*

About the size of an American Robin but with longer legs. Brown above, white below with two bold, black bands across breast. Except during nesting season, a bird of wetlands, often along coastal shores as wells as lake and river edges. For nesting, favors dry, barren areas where it places its three to four eggs in a shallow scrape on the ground. Because nesting areas include lawns and driveways, the Killdeer is familiar to many people. When a human, dog, cat, or wild predator approaches nest or young, female Killdeer will drag her wing as if it is broken, an attempt to lure the predator away. **Voice**: Loud, piercing *kill-deerrr, kill-deerrr.* **Length**: 10.5" (27 cm)

SPOTTED SANDPIPER *(Actitis macularius)*

Smaller than the Killdeer, the Spotted Sandpiper is exclusively found along shores of lakes, rivers, or coasts. During spring and summer when we are most likely to see it, note characteristic black spots on white breast and belly. Has distinctive behavior of bobbing its tail up and down constantly as it walks; in flight, snaps its wings in a stiff flutter. **Voice**: When startled, gives a high-pitched *weet,* sometimes whistled in pairs. **Length**: 7.5" (19 cm)

SOLITARY SANDPIPER *(Tringa solitaria)*

Larger and longer-legged than Spotted Sandpiper. This species is regularly seen in Maine during spring and fall migration on its way north and south to and from its breeding grounds in the boreal forests of Canada and Alaska. White spots on back; streaks (rather than spots) on breast separate it from Spotted Sandpiper. Bold eyering and short, olive-colored legs distinguish it from similar-looking Lesser Yellowlegs (not shown). Like its name suggests, the Solitary Sandpiper is most likely to be seen by itself rather than in large flocks like most sandpipers. Most likely to be seen along the edges of small inland ponds and other small wetlands. **Voice**: A piercing *pea-weet,* higher pitched than otherwise similar call of Spotted Sandpiper. **Length**: 8.5" (22 cm)

AMERICAN WOODCOCK *(Scolopax minor)*

One of the first signs of spring in Maine is when the loud nasal *peent* call of the newly returned male American Woodcock is heard at dusk from a brushy field as it attempts to lure a female. A chunky, short-legged bird with an unusually long bill, it is more often heard than seen. When well seen, look for the orangey underparts, black squares on back of head and nape, and large eye, which distinguish it from similarly shaped Wilson's Snipe that can occur along pond edges. American Woodcock are extremely secretive; their plumage pattern allows them to blend in well on the ground—they will remain still until almost stepped on, when they suddenly take flight. Gives elaborate flight display from open fields at dusk during breeding season, including wing twittering and harsh buzzy *peent* calls. Typically, several males can be heard (and sometimes seen) simultaneously during flight displays. Birds begin displaying in early March. **Voice**: A nasal *peent,* often heard at dusk, but only during breeding season. **Length**: 11" (28 cm)

KILLDEER

SPOTTED SANDPIPER

SOLITARY SANDPIPER

AMERICAN WOODCOCK

Wilson's Snipe for comparison

MOURNING DOVE *(Zenaida macroura)*

Medium-sized bird with a long, pointed tail. Small head, short legs. Body is light brown; tail has white outer edges. Easily identified and common at bird feeders throughout the state, feeding on platform feeders or on the ground. **Voice**: A plaintive *ooOOoo, oo, oo, oo.* **Length**: 12" (31 cm)

BELTED KINGFISHER *(Ceryle alcyon)*

Commonly seen on snags or plunging for fish near the edges of lakes, ponds, and rivers. Medium sized, with gray-blue head and back. White underneath. Males have a gray-blue breast band; females also have a red breast and flanks. Both have a shaggy crest. **Voice**: A loud rattling. **Length**: 13" (33 cm)

ROCK PIGEON *(Columba livia)*

The familiar pigeon of cities and towns. Most common form is gray with a black tail band, white rump, black bars on wings, and iridescent head, though most flocks contain individuals of a variety of colors and patterns from pure white to a reddish color. Although native to Eurasia, Rock Pigeons have been raised for food for thousands of years. Early settlers brought domesticated Rock Pigeons with them to North America, and escaped birds eventually become established around most cities. **Voice**: A low cooing. **Length**: 12.5" (32 cm)

BARRED OWL *(Strix varia)*

Large, dark-eyed owl lacking ear tufts. Hoots may be heard echoing across distant woods during late winter and early spring as the birds start their breeding season. Only other common resident large owl in Maine is Great Horned Owl which is larger, has prominent ear tufts, and yellow eyes. **Voice**: A distinctive *Who-cooks-for-you, who-cooks-for-YOU-all;* sometimes just *WHO-all.* **Length**: 20" (50 cm) **Wingspan**: 43" (110 cm)

RUBY-THROATED HUMMINGBIRD *(Archilochus colubris)*

The only hummingbird regularly seen in Maine—and eastern North America. Iridescent green above with distinctive, long, thin bill. Adult male shows a ruby-red throat in good lighting (in poor lighting, throat looks dark). Wings make a high, buzzy sound in flight, beating an amazing fifty-three times per second, making them almost invisible. Frequently hovers while feeding at nectar feeders or at colorful flowers. Displaying males do a pendulum-like display, with accompanying rising-and-falling buzzy sound. When foraging for food for nestlings, adults may be seen searching under eaves of houses for spiders and small insects to feed them. **Voice**: A scolding *chee-dit,* often given during chase. **Length**: 3.75" (10 cm)

CHIMNEY SWIFT *(Chaetura pelagica)*

A familiar summer bird of cities and towns, where it builds its nest of small sticks on the insides of chimneys. Affectionately called the "flying cigar" because of its gray cigar shape. Flies with narrow, pointed wings in a rapid flutter of wingbeats high in the air, constantly chattering as it searches for flying insects. In migration, large numbers sometimes descend at twilight into big chimneys where they roost for the night. Chimney Swifts are thought to winter in the Amazon Basin of South America, though there are few records of the species south of the U.S. **Voice**: Constant, rapid, high-pitched chattering—an obvious giveaway to its presence—is the background sound of summer in many cities and towns. **Length**: 5" (13 cm)

BELTED KINGFISHER

♀

♂

MOURNING DOVE

Great Horned Owl for comparison

ROCK PIGEON

BARRED OWL

♀

♂

RUBY-THROATED HUMMINGBIRD

CHIMNEY SWIFT

RED-BELLIED WOODPECKER *(Melanerpes carolinus)*

Once a very rare visitor to the state, Red-bellied Woodpeckers have now become uncommon but widespread year-round breeders following a major incursion in 2004. The species' most noticeable feature is not its red belly, which can rarely be seen, but rather its bright red nape contrasting with its zebra-striped back and warm tan undersides. **Voice**: A loud, nasal rolling *kwerr* rising up like a question. **Length**: 9.25"

YELLOW-BELLIED SAPSUCKER *(Sphyrapicus varius)*

Generally arriving in Maine in April and leaving by October, Yellow-bellied Sapsuckers are quiet and easily missed except during the spring breeding season. Then males are famous for their loud rapping on hollow trees and even on metal camp chimneys and other structures, often starting at daybreak, awakening unsuspecting human residents. The only Maine woodpecker with a bold, white vertical stripe on wings, a barred back, and yellowish underparts. Crown red, face black and white. Males have red chin and throat. Drumming is distinctive, starting fast then slowing. **Voice**: A nasal *eeyah.* **Length**: 8.5" (22 cm)

DOWNY WOODPECKER *(Picoides pubescens)*

Our most common year-round resident woodpecker. A small woodpecker with strongly marked black-and-white plumage. Males show red patch on back of head. Smaller but nearly identical in plumage pattern to Hairy Woodpecker. Look for the Downy Wood-pecker's proportionately smaller bill and dark barring on the outer tail feathers, a feature never shown by Hairy Woodpecker. **Voice**: Loud, sharp *peek,* and a whinny (higher in pitch than call of Hairy Woodpecker). **Length**: 6.75" (17 cm)

HAIRY WOODPECKER *(Picoides villosus)*

Medium-sized year-round resident woodpecker with strongly marked black-and-white plumage. Males show red patch on back of head. Identified from almost identical-looking Downy Woodpecker by larger size, proportionately larger bill, and lack of barring on outer tail feathers. Drumming is different from Downy Woodpecker—loud and fast, with abrupt beginning and end, reminiscent of a telephone ring. **Voice**: Loud, sharp *peek,* and a whinny (lower in pitch than call of Downy Woodpecker). **Length**: 9.25" (24 cm)

NORTHERN FLICKER *(Colaptes auratus)*

A common summer Maine resident and spring and fall migrant woodpecker. Rare winterer. Tan-colored, medium-sized with black barring on back and wings; black spots on breast and belly. Black crescent on chest. Gray crown with red crescent on nape. Male has black mustache stripe. In flight, look for its white rump and flashing yellow underwings. Unlike most other woodpeckers, Northern Flickers often can be found on the ground in open areas, where they specialize on feeding on ants. **Voice**: A loud series of *wick-wick-wick-wick-wick* calls, all on same pitch; also gives a *klee-ar.* **Length**: 12.5" (32 cm)

PILEATED WOODPECKER *(Dryocopus pileatus)*

A large, crow-sized woodpecker, the Pileated is Maine's largest and is virtually unmistakable. No other woodpecker in Maine has a black body with large, red crest and large bill. Male has a red mustache stripe; female's is black. Its large, rectangular-shaped excavation holes in rotting trees are distinctive. In flight, shows slow, deep, crow-like wingbeats. Territo-rial drumming, heard in spring, sounds like a distant jackhammer. **Voice**: A loud, echoing *kuk-kuk-kuk-kuk,* deeper and slower than that of Northern Flicker. **Length**: 16.5" (42 cm)

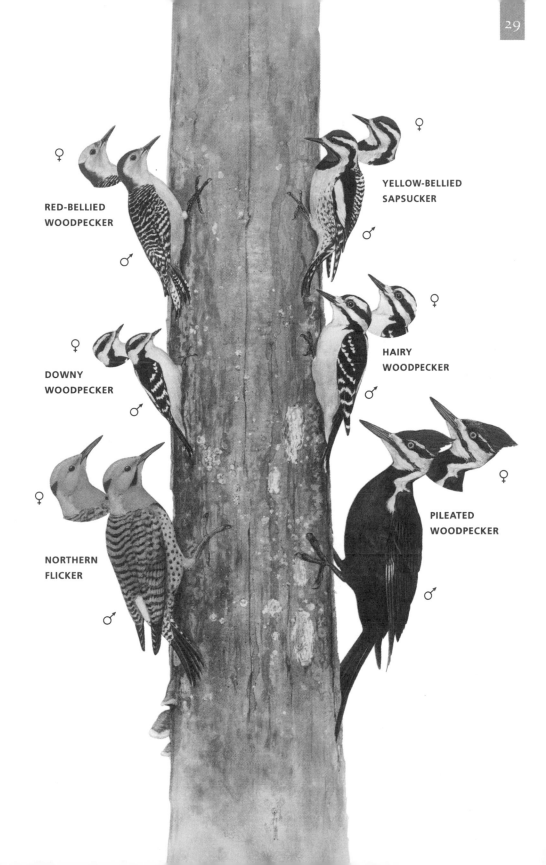

RED-BELLIED
WOODPECKER

♀

♂

YELLOW-BELLIED
SAPSUCKER

♀

♂

DOWNY
WOODPECKER

♀

♂

HAIRY
WOODPECKER

♀

♂

NORTHERN
FLICKER

♀

♂

PILEATED
WOODPECKER

♀

♂

GREAT CRESTED FLYCATCHER *(Myiarchus crinitus)*

A loud, boisterous summer-resident flycatcher of Maine's deciduous forests, almost as large as American Robin but slimmer. Yellow belly and undertail region contrasts with gray upper breast and throat; rather drab olive-gray upperparts. Note its two white wingbars and rufous in wings and tail. Unique among Maine's flycatchers, the Great Crested Flycatcher nests in tree cavities and nest boxes and often drapes shed snake skins from the nest hole. **Voice**: A loud, rough w*ee-eep.* **Length**: 8" (22 cm)

LEAST FLYCATCHER *(Empidonax minimus)*

Perhaps the most common and widespread of the four closely related small summer-resident flycatchers (the so-called *Empidonax* group) regularly found in Maine. All are similarly sized and show two wingbars and an eyering, but the species are separable based on subtle differences in shape, size, and coloration and their distinctive vocalizations. The Least Flycatcher is the common forest-dwelling species; others in this family are found in swampy shrublands and coniferous forests. Unlike the Eastern Phoebe, shows two distinctive wingbars, an eyering, and a pale base to bill. Belly and undertail sometimes pale yellow. Compared to other Maine *Empidonax* flycatchers, head is proportionately large and shows bold, white eyering. **Voice**: Best identified by its song— a quick, emphatic *che-bek;* also gives a sharp *wit* call. **Length**: 5.25" (13 cm)

EASTERN KINGBIRD *(Tyrannus tyrannus)*

A conspicuous, relatively large (robin-sized) summer-resident flycatcher with sharply contrasting black-and-white plumage—dark above, white below, with broad, white terminal band to black tail. Eastern Kingbirds are famous for attacking vastly larger birds like crows, hawks, and eagles. They nest along rivers and streams and in open areas where they noisily flutter out from high perches to catch insects. When a larger bird comes to within a few hundred yards of their nest, they will fly out in a characteristic fluttering flight to attack, and can sometimes be seen pecking and hitting the larger bird with their feet. **Voice**: A series of high-pitched sputtering notes followed by a downward *zeerr.* **Length**: 8.5" (22 cm)

EASTERN PHOEBE *(Sayornis phoebe)*

The first flycatcher to arrive back in Maine each spring, usually by late March or early April. Well known to Mainers for its habit of building its nests under the eaves of buildings. Dark gray-brown upperparts and tail, whitish underparts. No eyerings or distinctive wingbars. One of best identification features is its habit of bobbing its tail while perched watching for flying insects for food. The Eastern Phoebe is the first North American species to be banded—in the early 1800s by John James Audubon, who discovered what many Mainers would have guessed: that the same individuals return to and often nest right under the same eaves as they did the year before. **Voice**: Says its name—a distinctive and exuberant *fee-bwee.* **Length**: 7" (18 cm)

EASTERN WOOD-PEWEE *(Contopus virens)*

A common summer bird of Maine's oak and maple forests, often first noticed by its loud, piercing whistled *pee-err-wee* song. The Eastern Wood-Pewee is an inconspicuous olive-gray above, whitish below with white wingbars. Pale base of lower part of bill (lower mandible). Lacks eyering. Most readily confused with Eastern Phoebe, which lacks conspicuous wingbars, has an all-black bill, and is generally found in open areas or along forest edges. **Voice**: A slow, plaintive and piercing, whistled *pee-errr-wee.* **Length**: 6.25" (16 cm)

GREAT CRESTED FLYCATCHER

LEAST FLYCATCHER

EASTERN KINGBIRD

EASTERN PHOEBE

EASTERN
WOOD-PEWEE

TREE SWALLOW *(Tachycineta bicolor)*

One of the welcome signs of Maine spring is the arrival of the first Tree Swallows that sweep up from their southern U.S. wintering grounds on the first warm sustained south winds in March and April. Because they readily use backyard nest boxes, Tree Swallows are well known to Mainers. A relatively small songbird with dark, glossy greenish-blue upperparts that sharply contrast with its snowy-white underparts. First-year females usually show brown upperparts, as do juveniles. Tree Swallows are so-called "aerial insectivores," which means they catch their insect food while in flight and so are often seen over water bodies and adjoining habitats where insects are abundant. **Voice**: A chatty series of twitters. **Length**: 5.75" (15 cm)

BARN SWALLOW *(Hirundo rustica)*

As their name suggests, Barn Swallows are now almost exclusively found nesting on human-made structures—in and under the eaves of garages, sheds, houses, bridges, and other structures, including, of course, barns. This is the only small bird in Maine that has a long, deeply forked tail. Upperparts dark iridescent blue; underparts rusty. Throat and forehead reddish-brown. Like all swallows, most often seen flying low over open areas in search of insect prey; sometimes large groups will perch on telephone wires, especially in late summer when young have first started flying. **Voice**: A jumbled series of twittery notes, interspersed with a dry rattle—a familiar summer sound around farmyards with open barns. **Length**: 6.75" (13 cm)

BLUE-HEADED VIREO *(Vireo solitarius)*

One of Maine's early returning songbird migrants, coming back from its wintering grounds in the southern U.S. in April. A common but often overlooked nesting bird throughout the state. A relatively small songbird, its most striking features are its bold, white "spectacles" that contrast with a bluish-gray head. Back is greenish and shows bold, white wingbars; flanks are yellow. Prefers mixed coniferous-deciduous woods. In southern Maine, often found in hemlock groves along streams in oak and maple woodlands. **Voice**: Song is a series of phrases like an American Robin's, with each phrase ending in a down-slur or an upswing but very slow and nasal. **Length**: 5" (13 cm)

RED-EYED VIREO *(Vireo olivaceus)*

One of Maine's most widespread forest-inhabiting summer resident species, the Red-eyed Vireo is also one the most persistent singers and is more easily heard than seen. Even on hot summer days when other birds are quiet, the song of this species can be heard floating down from the canopy in its preferred habitat, deciduous forests. A small songbird with olive-green upperparts, white underparts, and gray crown. Look for its white eyebrow stripe bordered on bottom by a thin, dark eyestripe. Although it does have red eyes, they are hard to distinguish unless seen closely. Affectionately called "preacher bird" by some for its long-winded habit of singing all day. **Voice**: Song is a series of phrases like the familiar song of American Robin but higher and faster—a repeated series of rising-and-falling phrases sometimes rendered as *Here I am, over here, see me?* **Length**: 6" (15 cm)

TREE SWALLOW

BARN SWALLOW

BLUE-HEADED VIREO

RED-EYED VIREO

BLUE JAY *(Cyanocitta cristata)*

A familiar bird throughout Maine and a colorful, noisy, and a regular visitor to backyard feeders. Its striking blue-and-white plumage and large crest make it unmistakable, though the subtleties of its plumage are worth a closer look. Like most jays and crows, Blue Jays are very social birds and are often seen in groups. Maine's resident Blue Jays are joined in winter by migrants from farther north. During migration periods (especially in fall), these migrant birds can be seen in flocks passing overhead, sometimes in the hundreds. **Voice**: Most common vocalizations include a loud *jeer* or *jay* and clear whistled and gurgled notes. **Length**: 11" (28 cm)

AMERICAN CROW *(Corvus brachyrhynchos)*

Familiar to all, the American Crow is distinctive, all black, often seen in groups of two or more. Crows remain together in extended family units with younger birds often helping the breeding pair to feed and defend the young from predators. The virtually identical, slightly smaller Fish Crow is increasing but currently patchily distributed in summer in southern Maine (regular in Brunswick, Saco, Kennebunk, South Portland, and Rockland); it's identified only by its voice—a short, nasal, double-noted *uh-ha,* which itself can be hard to distinguish from the calls of American Crows, especially young individuals. The much larger Common Raven occurs throughout the state, though more commonly found farther north and east, distinguished by larger size, wedge-shaped tail, and proportionately larger bill that gives a raven's face a sloping appearance. **Voice**: A loud, harsh *caw, caw* and a variety of rattles and other sounds. **Length**: 17.5" **Wingspan**: 33.5–39" (85–100 cm)

BLUE JAY

juvenile

adult

AMERICAN CROW

juvenile

adult

BLACK-CAPPED CHICKADEE *(Poecile atricapillus)*

Maine's official state bird is a beloved and familiar sight and sound. A feisty little songbird, gray above and white below. White cheek and underparts contrast sharply with black cap and bib. Readily visits feeders, using bill to hammer seeds open. Also stashes seeds for later—a single bird can remember thousands of stashes. Also often seen scouring evergreens for insect eggs. Nests in holes in rotten trees and will use nest boxes. **Voice**: Says its name, *chick-a-dee-dee-dee-dee-dee;* also a sweetly whistled *hey-sweet* or *hey-sweet-ti,* which, unlike most bird songs, may be heard any time of year. **Length**: 5.25" (12 cm)

TUFTED TITMOUSE *(Baeolophus bicolor)*

Once an exceptionally rare bird in Maine, the Tufted Titmouse has spread northward into the state over the last four decades and is now common through the southern third of the state and has occurred north into Aroostook County and east to Washington County. A small, gray songbird with gray crest; large, dark eyes; black forehead; and buffy flanks. A regular year-round resident at backyard feeders, though usually in twos or threes—smaller numbers than in Black-capped Chickadee. **Voice**: A quick, clear whistled *pee-ter, pee-ter,* with emphasis on the first syllable; also, scolding notes and a harsh, chickadee-like *tsee-day-day-day.* **Length**: 6.25" (16 cm)

WHITE-BREASTED NUTHATCH *(Sitta carolinensis)*

A small, fairly tame songbird, the White-breasted Nuthatch is slightly larger than a Black-capped Chickadee. Known for its habit of walking up and down tree trunks, calling with its nasal voice. Note its black cap, white undersides, gray back and wings, and relatively long bill. A common year-round resident of Maine, though more common in the southern half of the state. It prefers deciduous forests and is often seen at feeders. Its smaller cousin, the Red-breasted Nuthatch, has an eyestripe and reddish breast, and prefers coniferous forests. **Voice**: Nasal *yank,* often repeated in a series. **Length**: 5.75" (15 cm)

BROWN CREEPER *(Certhia americana)*

Throughout much of the year, the only clue to the Brown Creeper's presence is its barely discernible thin, high-pitched *sreee* notes. A small, brown bird with a comparatively long and slightly curved bill. No other species has its habit of only ascending, never descending, the trunks of mature trees, which it does in search of food among crevices in the bark. When seen well, look for its brown upperparts speckled with white and contrasting white undersides; white eyebrow, long tail, and fine, down-curved bill. **Voice**: Its short, melodic whistled song is one of the first spring songs to be heard in woodland habitats, beginning in March, and is sometimes mistaken for an early returning warbler. **Length**: 5.25" (13 cm)

HOUSE WREN *(Troglodytes aedon)*

A common summer resident in southern and central Maine around backyards and open brushy areas. The House Wren repeats its song—a series of stuttering calls and trills—so regularly that it quickly becomes a background sound that may hardly be noticed. A small sprite of a bird often seen moving actively about a brushy tangle, giving glimpses of its short, often upturned tail, brown upperparts with barred wings and tail, light underparts, and slender, down-curved bill. The only species it is likely to be confused with is the smaller, darker Winter Wren, a coniferous forest-dwelling wren common in the northern two-thirds of the state. **Voice**: Along with its fast, bubbling song, also has a variety of harsh *churr* calls. **Length**: 4.75" (12 cm)

BLACK-CAPPED CHICKADEE

TUFTED TITMOUSE

WHITE-BREASTED NUTHATCH

BROWN CREEPER

HOUSE WREN

RUBY-CROWNED KINGLET *(Regulus calendula)*

A tiny bird with greenish upperparts, a bold, white eyering, white wingbar, and yellow edges to wing feathers and tail. Male's red crest only visible when bird is alarmed or agitated, at which time it raises it quickly, showing a flash of intense color. A useful identification feature is its habit of repeatedly flicking its wings as it darts about trees and shrubs in search of small insects. Although it nests in northern and eastern parts of Maine, the species is more regularly observed during spring and fall migration, when it is a common migrant throughout the state as it travels from and to its southeastern U.S. wintering grounds. **Voice**: Surprisingly loud and boisterous for such a small bird. Its song begins with several high, thin notes and accelerates into a rollicking series of whistles and trills. Also has a very characteristic but sometimes inconspicuous double-noted *chidit* call. **Length**: 4" (10 cm)

GOLDEN-CROWNED KINGLET *(Regulus satrapa)*

The Golden-crowned Kinglet is even slightly smaller than its tiny cousin, the Ruby-crowned Kinglet, with similarly patterned greenish upperparts, wings, and tail. Its head and face is very different, though, with black eyestripe, white eyebrow, and yellow (female) or reddish (male) cap bordered with black. Like Ruby-crowned Kinglet, it flicks its wings regularly. Found year-round in coniferous woods throughout Maine, though some migrate and may be found in a variety of habitats. **Voice**: Call is a thin, rapid *see-see-see* that can easily go undetected. Song begins with similar high *see* notes followed by slightly lower jumbled trill. **Length**: 3.75" (9.5 cm)

RUBY-CROWNED KINGLET

GOLDEN-CROWNED
KINGLET

EASTERN BLUEBIRD *(Sialia sialis)*

Seen in good light, the brilliant blue upperparts contrasting with the red breast make this small thrush—especially the more vivid males—one of Maine's most beautiful birds. As you drive Maine's country roads, look for Eastern Bluebirds perched on telephone wires, where their distinctive, hunched silhouette helps identify them even in poor light. Although primarily seen during breeding season, a few overwinter in Maine. Readily breeds in nest boxes in pasturelands, orchards, agricultural fields, and backyards near these habitats. Like Bald Eagles, bluebirds are a conservation success story. Their numbers crashed early in the twentieth century due, in part, to competition for tree holes from non-native European Starlings and House Sparrows introduced from Europe. Thanks to conservation efforts including bluebird trails (nest boxes monitored to keep out unwanted species), bluebird populations have recovered. **Voice**: A sweet, querulous *chur-lee*. **Length**: 7" (18 cm)

VEERY *(Catharus fuscescens)*

A relatively common bird of moist, mixed deciduous-coniferous forest, the Veery is more often heard than seen. Its descending flute-like song is a familiar sound from late May through early July, before it begins its migration to South America, where it spends the winter. Slightly smaller than an American Robin, the Veery is reddish-brown above, white below with indistinct spotting on breast. Pale eyering. **Voice**: Song is a rolling, rising-and-falling *veeeer-veeeer, veeeer-veeeer;* call is commonly a sharp, falling *veeer.* **Length**: 7" (18 cm)

WOOD THRUSH *(Hylocichla mustelina)*

Like other thrushes, a rather secretive bird to see, though its melancholy voice is familiar and much loved where it occurs in summer in Maine, primarily in the southern part of the state, in shrubby, young, deciduous forest habitats. Reddish-brown above with rustier head and large dark spots on breast. Bold, white eyering. **Voice**: Song is two or three flute-like phrases ending in a trill, often described as *ee-oh-lay;* call is a rapid-fire *pit-pit-pit-pit.* **Length**: 6.75" (17 cm)

AMERICAN ROBIN *(Turdus migratorius)*

A species familiar across Maine, and though many people consider it a sign of spring, flocks of robins often overwinter here, feeding on berries and other fruits. Commonly seen feeding on lawns in suburban neighborhoods but also occurs in fields, city parks, and elsewhere in a range of habitats, even forest openings in remote areas. Dark gray upperparts and brick-red breast make it distinctive. **Voice**: Song is a repeated *cheerio, cheery-up;* common calls include an urgent *chup-chup-chup* and a soft *pup.* **Length**: 10" (25 cm)

HERMIT THRUSH *(Catharus guttatus)*

The Hermit Thrush, like the other thrushes, has one of the most beloved voices of Maine's birds, clear and melancholy, often one of the last heard as evening settles, echoing out over lakes and coastal waters from its mixed and coniferous forest nesting habitat. A hardy thrush, a few linger even into the dead of winter, especially in south coastal Maine. Distinctive reddish tail noticeably contrasts with brownish upperparts. Underparts light-colored with dark spots on breast. White eyering. Often flicks tail upward when perched. **Voice**: Song is a single clear whistle followed by flute-like gurgle that trails off at the end; calls include a double *chuck* note and a nasal, rising *wee.* **Length**: 6.75" (17 cm)

GRAY CATBIRD *(Dumetella carolinensis)*

A medium-sized skulker of dense brush that is often more easily heard than seen, though singing males tend to perch on tops of bushes and small trees. Gray overall with dark cap and tail; rufous undertail coverts. **Voice**: A long series of chattery calls and squawks; also, the harsh, distinctive, catlike *mee-aw* that gives the bird its name. A relative of mockingbirds and thrashers, catbirds are able to imitate the songs of other birds, often forming them into their own song. **Length**: 8.5" (22 cm)

NORTHERN MOCKINGBIRD *(Mimus polyglottus)*

Although often thought of as a bird of the southern U.S., the Northern Mockingbird is now a regular but somewhat uncommon resident of southern Maine. Most often occurs in suburban areas, less frequently in rural settings. Once territory is established, the birds usually become a well-known presence by their loud songs, sung from exposed perches, including rooftops and telephone wires. Northern Mockingbirds are approximately robin-sized with a long tail with white sides, gray upperparts, and, in flight, bold, flashing white patches in the wings. **Voice**: As the name suggests, songs are a series of sometimes near-perfect imitations of a wide variety of other bird songs and even some human-made sounds (ringing telephones and truck back-up beeps, for example). Call is a harsh *smack* **Length**: 10.5" (26.7 cm)

EUROPEAN STARLING *(Sturnus vulgaris)*

Introduced from Europe into the U.S. in the 1890s, this species is now well established near cities and towns from Canada south into South America, including in Maine. . Distinctive blackish plumage is freckled with brown and white in winter and has a green sheen in summer. Sharp, relatively long bill (bright yellow in summer); short tail. Juveniles can begin appearing as early as late June and are sooty gray but with the same general shape as the adult. In the fall and winter, starlings gather into flocks sometimes numbering into the hundreds, even rarely into the thousands. Nests in tree cavities, nest boxes, and holes in buildings. Well known for its aggressive habit of evicting native cavity-nesting birds, including Eastern Bluebirds and Tree Swallows, from their nests so that they can use these nest sites themselves. **Voice**: Famous for its ability to do almost perfect imitations of many other bird songs, which it intersperses with harsher whistles and clicks. Except for a brief period in fall, sings much of the year. **Length**: 8.5" (21.5 cm)

CEDAR WAXWING *(Bombycilla cedrorum)*

A mysterious wandering species typically found in flocks, the largest appearing in the fall and early winter. A striking bird, with its distinctive crest and black mask. Brownish overall with grayish wings and yellowish belly; white undertail coverts. Note the bright yellow tail band. Inner flight feathers (secondaries) have red, waxy-looking tips from which the name "waxwing" is derived. Young birds are streaked on breast and lack red tips on secondary feathers. Through much of the year, waxwings feed largely on berries and fruits; in spring, often seen feeding on apple and cherry blossoms. In late summer, waxwings may be observed leaving high, exposed perches and flying out over water or open areas as they capture and eat flying insects. **Voice**: A soft, trilly, almost hissing high whistle. **Length**: 7.25" (8 cm)

♀

♂

EASTERN BLUEBIRD

VEERY

WOOD THRUSH

HERMIT THRUSH

AMERICAN ROBIN

NORTHERN MOCKINGBIRD

GRAY CATBIRD

EUROPEAN STARLING

juvenile

winter

summer

CEDAR WAXWING

OVENBIRD *(Seiurus aurocapilla)*

So named for its domed nests, which are built on the ground, making this species very vulnerable to neighborhood cats and other ground predators. Common during summer in Maine, but more easily heard than seen. Plain olive upperparts give it a thrush-like appearance, and beginning birders often mistake it for one. Note black-bordered orange crown stripe, white eyering. When seen, it is often walking on the ground with deliberate steps, though sings perched in branches 10–30 feet above ground. Prefers older, more mature deciduous forests. **Voice**: Loud *TEACHer-TEACHer-TEACHer-TEACHer.* **Length**: 6" (15 cm)

NORTHERN WATERTHRUSH *(Parkesia noveboracensis)*

A summer nesting resident of wet, swampy woodlands, this species is, like the Ovenbird, more easily heard than seen. Uniform brown upperparts; streaked, yellowish-white undersides and throat. Yellowish-white eyebrow tapers toward the rear, unlike in the rarer Louisiana Waterthrush that is found only locally in southern Maine, which shows an eyestripe that widens at the end. Stays low in thick, shrubby tangles; when spotted, note its habit of strongly pumping its tail. **Voice**: Variable, loud two- or three-part song uttered rapidly, sounding like *sweet-sweet-sweet, chew-chew-chew-chew, chewoo.* **Length**: 6" (15 cm)

COMMON YELLOWTHROAT *(Geothlypis trichas)*

Male has distinctive black "robber" mask bordered with white along top edge; yellow throat and unmarked olive upperparts. Female and immature similar but without black mask. A very common summer breeding bird throughout Maine. Yellowthroats prefer dense, brushy habitat in open areas such as old fields or along the edges of woods, streams, and marshes. Typically, they stay very low and try to remain hidden. **Voice**: Song is a persistent *witchety-witchety-witchety.* **Length**: 5" (13 cm)

WILSON'S WARBLER *(Cardellina pusilla)*

Striking black cap contrasts sharply with yellow-olive upperparts and strong yellow underparts, making this species fairly easy to identify. A Central American winterer, it usually arrives in Maine in mid-May. Most commonly seen during migration, but nests in summer in marshy areas from central Maine north. Even during migration, prefers brushy areas, often near water. **Voice**: Song is a stumbling, descending loose trill usually speeding up at the end. **Length**: 4.75" (12 cm)

OVENBIRD

NORTHERN WATERTHRUSH

COMMON YELLOWTHROAT

♀

♂

WILSON'S WARBLER

NASHVILLE WARBLER *(Oreothlypis ruficapilla)*

A regular breeder and common spring and fall migrant often arriving in southern Maine by late April and early May from its Central American wintering grounds. Named for Nashville, Tennessee, where it was first discovered, though it only occurs in that region for a short time during migration. Bright yellow underparts; greenish-yellow upperparts contrasting with gray "helmet." Narrow, white eyering; lacks wingbars. **Voice**: Song typically three two-part notes followed by loose trill *seebit-seebit-seebit, titititititi.* **Length**: 4.5" (11 cm)

NORTHERN PARULA *(Setophaga americana)*

Sometimes abundant during spring and fall migration, the Northern Parula is a relatively common summer breeding bird. A small, colorful warbler with a short tail and rather thin bill. Light blue tail, wings, and head. Greenish "saddle" on back. White wingbars, broken, white eyering, and yellow throat and upper breast divided by black and chestnut breast band in male. In females and immatures, breast band is much reduced or lacking altogether. White extending from belly to underside of tail. **Voice**: Most common and recognizable song is a high-pitched smooth, rising buzz with an abrupt ending, *zeeeeeee-up!* **Length**: 4.5" (13 cm)

YELLOW WARBLER *(Setophaga petechia)*

One of Maine's most widespread and easily observed warblers. Found in brushy areas throughout the state, especially near water. As its name suggests, this is the only warbler that is completely yellow, including its tail. In breeding plumage, the face is bright yellow, and the underparts show either red streaks (male) or dingy gray streaks (female and immature) on breast. Upperparts yellow-green. Usually arrives in early May in southern Maine; by mid-May its song can be heard ringing out from brushy areas throughout the state. **Voice**: Song is a high, fast-whistled *Sweet-sweet-sweet-I'm-so-sweet.* **Length**: 5" (13 cm)

CHESTNUT-SIDED WARBLER *(Setophaga pensylvanica)*

A bird of brushy edges and burned or cut-over forest filling in with blackberries and other shrubby growth. Unlike many warblers, its undersides are white instead of yellow. Note the distinctive chestnut sides (less extensive in female), the black line through eye and black "teardrop" in front of eye. Male has bright, lemon-yellow cap. Back is green-yellow spotted with dark gray; wings dark gray with two white wingbars. Female coloration duller overall than male. **Voice**: Song a whistled *Pleased-pleased-pleased-to-meetcha.* **Length**: 5" (13 cm)

NASHVILLE WARBLER

NORTHERN PARULA

YELLOW WARBLER

CHESTNUT-SIDED WARBLER

MAGNOLIA WARBLER *(Setophaga magnolia)*

Breast streaks across the yellow breast create a distinctive black "necklace" below a bright yellow throat. Grayish-blue upperparts with large, white wingbar, white eyebrow, and black mask. A common summer nesting bird throughout Maine in its preferred habitat of moist mixed coniferous-deciduous woods. In spring and fall migration, often one of the most common warbler species, especially along coastal locations. **Voice**: Song is a sweet whistled *weeta-weeta-weeta-weeteeo*. **Length**: 5" (13 cm)

BLACK-THROATED BLUE WARBLER *(Setophaga caerulescens)*

Male is dark blue above with black throat extending along sides of breast to flanks, contrasting with immaculate white undersides. Both sexes have large, white square in folded wing (sometimes described as white "handkerchief"). Females are rather drab brown-gray on upperparts and dingy underneath, but look for white "handkerchief" on wing. Maine could probably be considered the world capital for Black-throated Blue Warblers, as the state supports a large proportion of the species' total world population. Prefers deciduous or mixed forest with shrubby understory, most abundant in beech-dominated forests. **Voice**: Buzzy, drawling *zur zur zur zree,* rising at the end, often described as *I'm so la-zee*. **Length**: 5.25" (13 cm)

YELLOW-RUMPED WARBLER *(Setophaga coronata)*

Maine's earliest returning warbler each spring, arriving before most trees have leafed out. Yellow rump is one of the most obvious features, inspiring the nickname "butter butts" among birders, though other warblers also have that field mark. Yellow-rumps also have two white wingbars, a "necklace" of dark streaks on upper breast, and bluish (male) or gray (female) upperparts, white throat, and yellow patch on top of head. Two bright yellow patches on sides of breast like a pair of oncoming car headlights when viewed from the front. These early arrivals are also the last warblers to leave Maine, and a few have been known to overwinter, mostly along the coast where they often feed on bayberry. **Voice**: A weak, loose trill or series of whistled notes. **Length**: 5.5" (14 cm)

BLACK-THROATED GREEN WARBLER *(Setophaga virens)*

One of Maine's most common summer breeding warblers, usually appearing by late April, especially from the midcoast area north, occurring almost exclusively in areas with spruce, fir, and hemlock trees. Often stays high in trees, making it difficult to see. Bright yellow face offset by black throat; greenish-olive cap and back are distinctive. Two white wingbars. **Voice**: Memorable song is a high, buzzy *zee zee zee zoo zee* dropping on the *zoo* and rising on the last note. **Length**: 5" (13 cm)

BLACKBURNIAN WARBLER *(Setophaga fusca)*

An unmistakable, strikingly beautiful warbler, though typically hard to see because it often feeds at the very tops of tall coniferous trees. Many people are surprised to learn that such a bright, tropical-looking bird is a common and widespread nester in Maine, especially from midcoast Maine north. Usually returns from its South American wintering grounds in May. Bright orange (male) or yellow-orange (female) face and throat with black "arrowhead" behind eye. Two white wingbars, often extending into a single large, bold, white wingbar in males. **Voice**: Extremely high and variable song; most straightforward and easiest to identify often described as *seebit seebit seebit ti-ti-ti-ti teeeee,* with very characteristic note on the end that seems to rise steeply in pitch until it goes beyond human hearing range. **Length**: 5" (13 cm)

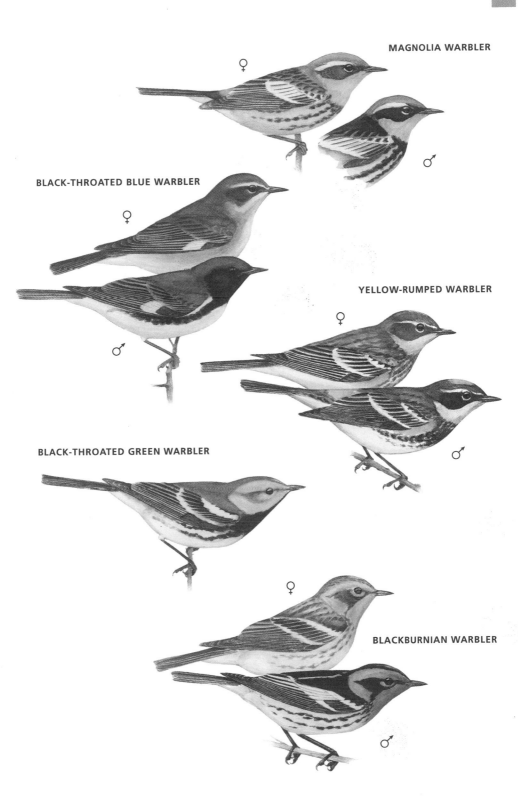

MAGNOLIA WARBLER

♀

♂

BLACK-THROATED BLUE WARBLER

♀

♂

YELLOW-RUMPED WARBLER

♀

♂

BLACK-THROATED GREEN WARBLER

♀

BLACKBURNIAN WARBLER

♂

PALM WARBLER *(Setophaga palmarum)*

One of the first warblers to return to Maine in the spring and among the last to leave in the fall, as it is a common wintering species in the southeastern U.S. Often pumps its tail, a quick clue to its identification. In spring, has yellow underparts with rufous breast streaks and rufous cap. Some have far less yellow and lack rufous streaks. Fall plumage is much drabber. During migration, look for it in shrubby areas, often near water. Breeds exclusively in open sphagnum bogs from Maine north through Canada. **Voice**: Song is an unremarkable loose trill. **Length**: 5.5" (14 cm)

BLACKPOLL WARBLER *(Setophaga striata)*

Males have a very distinctive black cap (poll), offset by a strong, white cheek. Females and fall-plumaged birds are less striking but have two white wingbars, light streaking on sides of breast, white undertail coverts, streaked back, and yellowish legs and feet. This species winters in South America, and its long-distance migration leads to their being among the latest spring arrivals in Maine—typically in late May, when large numbers flood through in a short time on their way north. In fall, its presence is not as obvious, but large numbers migrate through from mid-September through October. Blackpoll Warblers nest on mountaintops in Maine and in lower-elevation areas in northern and western Maine, with the extensive breeding range reaching through boreal portions of Canada to Alaska. **Voice**: Song is a very distinctive series of exceptionally high, thin notes that gradually get louder, then softer, very similar to the squeaking noise from car brakes. **Length**: 5.5" (13 cm)

BLACK-AND-WHITE WARBLER *(Mniotilta varia)*

A common forest-breeding bird throughout the state and one of the earliest migrants to arrive, usually in late April and early May. Its striped black-and-white "zebra" plumage is distinctive. Females and immatures lack black throat. Similar to Blackpoll Warbler but has white eyebrows, a white central crown stripe, black legs, and a longer, more down-curved bill. Has a characteristic nuthatch-like behavior of walking up and down tree trunks and limbs. A bird that's easy to see, if time is taken to track down the distinctive song. **Voice**: Song is a series of two-syllable repeated notes, sounding like *weesa-weesa-weesa-weesa*. **Length**: 5.25" (13 cm)

AMERICAN REDSTART *(Setophaga ruticilla)*

Adult males have striking coloration, with black back, head, throat, and upper breast; bright orange spots on breast, orange wingbars, and orange in tail. Females and first-year males have yellow instead of orange and have greenish-gray back and wings and gray head. Common, easily seen spring and fall migrant and breeder in young forested areas with understory. **Voice**: Several song types, but perhaps most common is a series of whistled notes with a down-slurred, accented ending, sounding like *see-see-see-seeo*. **Length**: 5.25" (13 cm)

PALM WARBLER

BLACK-AND-WHITE WARBLER

♂

BLACKPOLL WARBLER

♂

♀

♀

juvenile

♂

AMERICAN REDSTART

♀

♂

first spring

EASTERN TOWHEE *(Pipilo erythrophthalmus)*

A robin-sized, ground-foraging skulker that rambunctiously scratches the ground with both feet to uncover food items under leaves and debris. Upperparts, throat, and upper breast black (male) or chocolate-brown (female) with rusty flanks and white marks in wing. White corners of tail visible during flight. Towhees prefer brushy areas along field edges and under power lines where they can stay remarkably well hidden, though singing males often sit on exposed perches where they can be easily viewed. A summer nesting species in Maine, usually returning by April and leaving by October. **Voice**: Song is a whistled *Sip-your-teeeee*. Call note says its name, *towhee*, rising on *hee*. **Length**: 8.5" (21.5 cm)

AMERICAN TREE SPARROW *(Spizella arborea)*

Reddish cap, black-and-yellow bill, dark spot on breast, and white wingbars. Winter visitor from its northern Canadian and Alaskan breeding range. Sometimes confused with summer resident Chipping Sparrow, but American Tree Sparrow usually arrives in November and is gone by March, before most Chipping Sparrows arrive. Keep an eye out for it at feeders and in areas of low brush and weeds. **Voice**: Most often heard sounds in winter are sweet tinkling call notes, sometimes given by many birds calling together in a flock. **Length**: 6.25" (16 cm)

FIELD SPARROW *(Spizella pusilla)*

Another sparrow with a reddish cap, but note unmarked breast; narrow, white eyering; and pink bill. Summer resident that nests in overgrown fields in southern Maine. During spring and fall, look for it in mixed flocks of migrant sparrows in weedy areas. Rarely, a few persist into winter at feeders in southern Maine. **Voice**: Song is a series of down-slurred whistles, increasing in speed until ending in a trill. **Length**: 5.75" (14.5 cm)

CHIPPING SPARROW *(Spizella passerina)*

The smallest of Maine's common, reddish-capped sparrows. Unmarked breast, black bill, white wingbars. A summer-resident sparrow of yards, lawns, and parks (though also occurs in many natural habitats as well) throughout Maine. Common breeder, also readily seen in spring and fall as a migrant. **Voice**: Fast, extended trill repeated all day throughout the summer breeding season. **Length**: 5.5" (14 cm)

FOX SPARROW *(Passerella iliaca)*

One of the largest of the sparrows. Formerly known as a migrant only in Maine but now known to breed in scattered locations in northern Maine. Rufous overall with contrasting gray on head and neck, thick rufous streaks on breast, and gray and rufous streaking on back. The species is most often seen during migration but is relatively common for only a short time each spring and fall. Often seen scratching vigorously among leaves in brushy thickets but can also be found at feeders. **Voice**: Song is rich, whistled notes, fuller and more musical than that of most other sparrows. **Length**: 7" (18 cm)

EASTERN TOWHEE

♀

♂

AMERICAN TREE SPARROW

CHIPPING SPARROW

FIELD SPARROW

FOX SPARROW

SONG SPARROW *(Melospiza melodia)*

One of the most common sparrows, with a song that is a welcome sign of spring. Brown overall, streaked breast with central steaks running together to a dark breast spot. Breeds commonly in brushy areas throughout Maine, from suburban yards to wild areas. Also a common spring and fall migrant; uncommon winter visitor to feeders or brushy areas in coastal Maine. **Voice**: Song variable, but usually begins with three or four clear notes followed by a series of trills and whistles. **Length**: 6.25" (16 cm)

SWAMP SPARROW *(Melospiza georgiana)*

Rusty cap and wings contrasting with gray face. Whitish, unmarked throat; gray across upper breast, often with some blurry streaking. Breeds in open marshy areas throughout Maine, sometimes seen singing from the top of a cattail or alder shrub. During spring and fall migration, may be found in flocks with other sparrows feeding in weedy areas. A rare overwinterer. **Voice**: A loose, sweet trill. **Length**: 5.75" (14.5 cm)

WHITE-CROWNED SPARROW *(Zonotrichia leucophrys)*

Bold, black-and-white crown stripes, pinkish bill, unstreaked breast, and white wingbars. Often raises its crown feathers to give puffy-headed profile. Breeds in Canada, so migrants pass through Maine during a few weeks in May and much of October. Look for it in brushy areas and at feeders. **Voice**: Song begins with several clear whistled notes, followed by a series of buzzy trills, sometimes compared to *Oh, jam-jam feed-the-ants*. **Length**: 7" (18 cm)

WHITE-THROATED SPARROW *(Zonotrichia albicollis)*

For many, the song of the White-throated Sparrow is the quintessential summer sound of the North. Its namesake white throat contrasts with a grayish breast and sides of face. Adults occur in two color forms, one with white-and-black striped head, the other with tan-and-dark-brown head. Common spring and fall migrant; uncommon in winter, but in mild winters can be fairly common. Breeds through much of the state, most commonly in all but the southern portion, preferring brushy openings in mixed and coniferous forest. **Voice**: Clear whistled song famously described as *Old sam peabody-peabody-peabody*. **Length**: 6.75" (17 cm)

DARK-EYED JUNCO *(Junco hyemalis)*

A common, easily recognizable bird found throughout Maine. Upperparts, throat, and breast dark gray (males) or lighter gray (females and immatures) with pink bill. White sides of tail distinctive when birds flush, as often happens when flocks congregate along roadsides during migration. A widespread breeder throughout the state, an abundant spring and fall migrant, and a regular visitor to feeders in winter. Old-timers sometimes refer to juncos as "snowbirds" because of their tendency to arrive around backyards in late fall or early winter. **Voice**: Song is a high, fast trill, often difficult to distinguish from song of Chipping Sparrow. **Length**: 6.25" (16 cm)

SONG SPARROW

SWAMP SPARROW

WHITE-CROWNED SPARROW

WHITE-THROATED SPARROW

tan-striped form

white-striped form

♀

DARK-EYED JUNCO

♂

SCARLET TANAGER *(Piranga olivacea)*

A bird that delights all who see it. Males have a bright red, almost luminous body with strikingly contrasting black wings and tail. Female is drab by comparison, with greenish upperparts and yellow-green underparts but with characteristic tanager shape and bill. During breeding season, prefers upper, denser reaches of mature deciduous trees, making it easier to hear than see. Watch for them to arrive back in Maine in mid to late May from their South American wintering grounds. People are often surprised to know that the Scarlet Tanager is a relatively common bird, nesting in deciduous and mixed forests throughout most of the state. **Voice**: Song is often described as "a robin singing with a sore throat." **Length**: 7" (18 cm)

NORTHERN CARDINAL *(Cardinalis cardinalis)*

A favorite among favorites, for its colorful appearance as well as its cheery song. Male's all-red plumage, crested head, and black face set it apart from all other birds. Female's plumage is more subtle but also colorful—gray-brown with reddish wings, tail, and crest. The female's bill is reddish like the male's. Though fairly rare until about thirty or forty years ago, cardinals are now common in brushy areas of residential neighborhoods and a regular at bird feeders in all but northern portions of Maine. Unlike the females of most North American songbird species, female cardinals sing, sometimes from the nest, perhaps signaling to the male that it's time to bring her a meal. **Voice**: Well-known series of slurred, piercing whistles, *cheer-cheer-cheer* and *birdy-birdy-birdy.* **Length**: 8.75" (22 cm)

ROSE-BREASTED GROSBEAK *(Pheucticus ludovicianus)*

Beginning birders often think they're seeing a bird with a chest injury when they see a male Rose-breasted Grosbeak for the first time, as the bright red on its breast could be mistaken for blood, especially given its contrast with the white belly and sides and black upperparts. Females are brown with bold eyebrow stripe and streaked breast; large, ivory-colored bill, like the male's. Both sexes have white wingbars. A common migrant and summer nesting bird that prefers second-growth deciduous forests. In spring, attracted to backyards by orange slices or jelly. **Voice**: Song sounds like a syrupy-sweet robin. Call note is unmistakable, like the squeaking sound of a sneaker stopping suddenly on a gym floor. **Length**: 8" (20 cm)

INDIGO BUNTING *(Passerina cyanea)*

In good light, the breeding-plumaged male Indigo Bunting is a spectacular sight—luminous blue with silvery bill. Females and immatures have similar shape but are brown overall, sometimes with a bluish tinge to wings and tail. A relatively common summer breeding species that prefers open shrubby areas along field edges, large forest openings, and areas under power lines, though sometimes appears at feeders in spring. **Voice**: Song is finch-like except that each note is sung twice, *cheet-cheet, teer-teer, sweet-sweet.* **Length**: 5.5" (14 cm)

SCARLET TANAGER

♀

♂

NORTHERN CARDINAL

♀

♂

ROSE-BREASTED GROSBEAK

♀

♂

INDIGO BUNTING

♀

♂

BROWN-HEADED COWBIRD *(Molothrus ater)*

Male is readily identified by black body and brown head; female and immature are brownish gray but maintain characteristic shape and stout bill. Common spring and fall migrant and summer resident across the state. Usually occurs in small flocks, frequently at backyard feeders and agricultural areas, but can be seen almost anywhere. The Brown-headed Cowbird does not build nests or raise its own young but is a "nest parasite"; the female lays its single egg in another bird's nest. The parasitized species often does not recognize that an egg has been added and raises the cowbird as its own. **Voice**: Male's song is a liquid whistle; calls include liquid rattle and high, thin *see-tee*. **Length**: 7.5" (19 cm)

RUSTY BLACKBIRD *(Euphagus carolinus)*

Once a very common migrant and summer resident of Maine, the Rusty Blackbird has sadly declined by more than 90 percent over the last half-century, the reason for which has yet to be determined. Despite the decline, small numbers still are seen in Maine during migration and sometimes in winter at backyard feeders. Scattered breeding locations are also still known near streams and ponds in wet, coniferous woodlands in northern Maine. In fall, shows rusty brown-edged feathers on wings and back; yellow eye and pale eyebrow line. In spring, male is black; female is grayish. **Voice**: Song likened to the sound of a rusty hinge, sometimes described as *Here-I-bee*. **Length**: 9" (23 cm)

RED-WINGED BLACKBIRD *(Agelaius phoeniceus)*

One of the earliest migrants to return in spring, with small flocks often arriving in late February. It is not uncommon in March to see male Red-winged Blackbirds singing from treetops beside still hard-frozen lakes and marshes. Male is jet black with bold red shoulder edged with pale yellow. Female is brown with streaked breast. Large flocks occur in migration near wet areas and at backyard feeders. Common breeder in wetlands throughout the state. **Voice**: Song is a liquid *onk-er-lee,* rising on the last syllable. **Length**: 9" (23 cm)

BALTIMORE ORIOLE *(Icterus galbula)*

The arrival of this tropical-looking species, usually in May, is a cheery sight—the males showing bright orange underparts, shoulder, and rump contrasting with black head and back. Females are duller orange, with black replaced by mottled browns and lacking orange shoulders. Common in spring and fall migration, often lured to backyard feeders in spring with orange slices or jelly. Nests throughout the state, preferring second-growth deciduous forest edges near water, though becomes very secretive soon after nesting. **Voice**: Song is a loud piping whistle. **Length**: 8" (20 cm)

COMMON GRACKLE *(Quiscalus quiscula)*

A common and familiar bird that frequents backyard feeders during spring and fall migration and is regularly seen in wet woodlands and shrubby areas near lakes, ponds, and streams. A bit larger than a robin, with a rather large, imposing bill. Adult shows yellow eye, bronzy-black body with purplish sheen to head, long tail. Juveniles are sooty brown; eye is dark. Very common throughout Maine from early spring through fall. During migration, often seen with flocks of Red-winged Blackbirds. **Voice**: Song is harsh, squeaky, grating notes. **Length**: 12.5" (32 cm)

BROWN-HEADED COWBIRD

♂

♀

RUSTY BLACKBIRD

fall plumage

BALTIMORE ORIOLE

♀

♂

♂

♂

♀

RED-WINGED BLACKBIRD

♀

COMMON GRACKLE

♂

PINE SISKIN *(Spinus pinus)*

Small, dark brown-streaked finch, slightly smaller than a chickadee, with yellow in wings and tail and sharp, pointed bill. Although a year-round resident in coniferous forests in northern and eastern Maine where it nests in summer, known to most Mainers as a winter visitor to backyard feeders. On average, large numbers migrate south from Canada to winter in the eastern U.S. every two years; in those years, backyard feeders can have dozens, even occasionally hundreds of Pine Siskins. Sometimes occurs in winter at feeders with flocks of American Goldfinches and/or Common Redpolls. **Voice**: Most distinctive sound is a shrill rising *shreee*. **Length**: 5" (13 cm)

AMERICAN GOLDFINCH *(Spinus tristis)*

One of the most familiar—and loved—birds in North America. Sometimes referred to as "wild canaries." Breeding-plumaged male is bright yellow with black forehead and tail, black wings with white wingbars. Females and non-breeding-plumaged males are a drab greenish-yellow and lack black forehead. In direct flight, regularly dips down and rises up again. Common year-round throughout Maine, though in winter most often seen at feeders, sometimes in flocks of dozens. In spring, summer, and fall, watch for them in open, brushy areas; in spring, large flocks often feed near tops of deciduous trees in a loud, twittering chorus. **Voice**: A variety of calls and an extended warbling song given in late summer, though most memorable is their distinctive flight call—a sweet, clear, repeated *potato-chip*. **Length**: 5" (13 cm)

EVENING GROSBEAK *(Coccothraustes vespertinus)*

The massive, ivory-colored bill and bright yellow eyebrow of the males of this nearly robin-sized finch are distinctive. Both sexes have black wings and tail with large, white patch in wing. Male shows bright yellow belly and shoulder; dark brownish upper back and head. Female is grayish. Although small numbers breed in spruce-fir forest habitats in northern and eastern Maine (sometimes midcoast Maine as well), the species was formerly most well-known as a common winter visitor to backyard feeders. It has become increasingly uncommon even as a winter feeder bird, as its overall population has shown steady decline over the last thirty years (the reason for the decline has not been determined). Still occasionally seen at feeders. **Voice**: Call heard most often in winter is a loud, ringing *keeyou*. **Length**: 8" (20 cm)

COMMON REDPOLL *(Acanthis flammea)*

A small, short-legged finch with dark face encircling a small, sharply pointed yellow bill. Red forehead, white wingbars, and streaked upperparts and breast. Dark, slightly forked tail. Male with reddish-tinged breast. Breeds in far northern Canada and Alaska with large numbers appearing south into the northern U.S. every two years on average, often frequenting backyard feeders here in Maine in winter. The much rarer Hoary Redpoll is difficult to distinguish but is paler with unstreaked rump and undertail coverts and a shorter bill that gives its face a pushed-in look. **Voice**: Call, heard most often in winter, is a hard *chit-chit-chit*. **Length**: 5.25" (13.5 cm)

PINE SISKIN

♀

♂

♀

AMERICAN GOLDFINCH

♂

♀

EVENING GROSBEAK

♂

Hoary Redpoll

Common Redpoll

♀

COMMON REDPOLL

♂

HOUSE FINCH *(Carpodacus mexicanus)*

A common bird of towns and cities, and a regular visitor to backyard bird feeders. Male has reddish breast and head, brown back, wings, and tail. Female is brown with streaked breast; lacks white eyebrow of similar-looking female Purple Finch. Brown streaks on flanks and more curved bill distinguish male from Purple Finch. Originally introduced into the New York City area in the late 1940s from its original western U.S. range, it has now spread throughout most of eastern U.S. and southern Canada, where it is especially common in urban and suburban areas. **Voice**: Call a barking, yodeling *chow.* Song an extended musical warbling interspersed with occasional harsh, descending notes. **Length**: 6" (15 cm)

PURPLE FINCH *(Carpodacus purpureus)*

Similar in general appearance to House Finch, but both male and female show a more deeply forked tail and broader-based, less curved bill. Note also the male Purple Finch's more rosy-purple color as compared to the reddish tones in the male House Finch. Lack of brown flank streaking in Purple Finch also distinctive, though in bad light reddish flank streaks can look brownish. Female and immature Purple Finch brown with bold, white eyebrow in contrast with the plain, unmarked face of the female House Finch. Breeds throughout Maine but perhaps most frequently seen when it visits backyard feeders during migration and in winter. **Voice**: Characteristic call is a sharp *pik.* Song is a rich, extended warble usually lacking harsh notes characteristic of House Finch song. Also a surprisingly accomplished mimic, incorporating songs and calls from many other species into their own rambling songs. **Length**: 6.25" (16 cm)

HOUSE SPARROW *(Passer domesticus)*

Introduced to New York City from Europe in the late 1800s, the House Sparrow has successfully colonized from Canada south to Chile. Male has brown upperparts; bold, white wingbar, black throat and eye patch set against grayish-white cheek. Gray crown. Female has pinkish bill; tan-brown upperparts, drab gray underparts. Generally now found most commonly in cities and towns, where they frequent feeders or places where they can scavenge for food, including the parking lots of fast-food restaurants. Also found around farms with livestock, where spilled grain provides a food source. **Voice**: Calls include a loud *cheeup* and a rough, loose rattle. **Length**: 6.25" (16 cm)

HOUSE FINCH

♂

♀

PURPLE FINCH

♀

♂

HOUSE SPARROW

♀

♂

BIRDING IN MAINE

Birds can be seen and heard all around us in our everyday lives, if we look and listen for them. It's also fun, though, to visit places known as "birding hot spots." Nestled between the Canadian North and the mid-Atlantic states of the U.S., Maine has an interesting variety of hot spots. Some of these locations are also great places for a family hike, a quiet kayak or canoe excursion, sightseeing, or other activity.

We provide for you here some of the better known and easily accessed birding hot spots throughout the state. All of these places can be easily located in the *Maine Gazetteer* or by searching online, and all are worth visiting.

MARGINAL WAY, OGUNQUIT

Located in one of Maine's premier tourist towns near the southern tip of the state, this popular walking path offers fantastic views of Maine's rocky coast—and the birds that live there. Marginal Way can draw hundreds of visitors on a sunny day from late spring through early winter, and many folks enjoy the walk even in winter, if snow and ice conditions allow. Year-round, this is a great place to see Common Eiders and Herring and Great Blacked-backed gulls. You'll see Double-crested Cormorants in spring, summer, and fall. In the brushy areas along the walk, Yellow Warblers are abundant. During migration, keep an eye and ear out for White-throated Sparrows.

EAST POINT SANCTUARY, BIDDEFORD POOL

Jutting out into the Atlantic Ocean, Eastern Point Sanctuary at Biddeford Pool is not only a favorite walk for families but an incredibly good spot to find birds. Common Eiders are always present along the shore, and in summer you are often provided easy views of the downy chicks being tended to by the females. Herring and Great Black-backed gulls abound. During spring and fall migration, watch for a wide variety of landbirds, including warblers (Yellow-rumps are especially plentiful in fall) and sparrows (White-throated and White-crowned), from the trails. In summer, this is a good spot for Eastern Towhee and Gray Catbird. Along the shore, look for a variety of ducks, shorebirds, and waterbirds. During migration, Biddeford Pool Beach and the pool itself (marshy area along the road that parallels the beach) often have impressive numbers and varieties of sandpipers and plovers, and sometimes lots of American Black Ducks.

KENNEBUNK PLAINS

Situated right on Route 99 between Kennebunk and Sanford and only about a ten-minute drive from Route 1 or the Maine Turnpike (I-95), Kennebunk Plains is a birding gem. Park in the parking lot and enjoy the miles of easy-to-walk sandy roads that wind through a 500-acre blueberry barren. (In July and into August, bring your bowls and pick berries while you enjoy the birds around you.) Be sure to visit the area on McGuire Road, off Route 99; there is less traffic on this mostly dirt road, so you can hear the birds much more easily. From May through August, the Plains is full of singing birds, including many rare species such as Upland Sandpiper, Prairie Warbler, Grasshopper Sparrow, and,

at night, Whip-poor-wills. Some of Maine's better-known birds also reside here, among them Eastern Bluebirds, Eastern Towhees, and Field Sparrows. Along the edges, look and listen for the beautiful Indigo Bunting, and Pine, Chestnut-sided, and Yellow warblers. Keep your eyes on the sky for Turkey Vultures and hawks, and watch the wires for American Kestrels.

SCARBOROUGH MARSH
This locale is one of the most birded sites in Maine for its wide variety of shorebirds. There are a number of different access points. The most widely used is probably the raised Eastern Trail that bisects the middle of the marsh and is often accessed by parking on the south side of the marsh just off Pine Point Road. Another favorite access spot is behind the Pelreco complex on the right after the railroad overpass on Pine Point Road. It's always worth a stop down at Pine Point itself, near the Pine Point Fishermen's Co-op. Here you have a view of the estuary where the marsh drains into the ocean. Great Blue Herons and other herons and wading birds feed here, as do Tree and Barn swallows and a host of other familiar birds. Red-winged Blackbirds, American Black Ducks, and Mallards breed in the marsh, and Belted Kingfishers may also make an appearance.

PORTLAND AREA
In and around Maine's largest city awaits a treasure-trove of great places to look for birds. One of the most well-known and frequently visited sites during spring migration is Evergreen Cemetery. Not only does this birding hot spot often host just about every regular migrant songbird that is possible to see in Maine, but it has also been visited by rarities from more southerly climes, such as Hooded Warbler and Worm-eating Warbler. Back Cove, with its conveniently wide walking and biking path, offers views of an impressive variety of waterfowl species and other birds that rely on wetlands, especially in migration. The area near the ball fields on the south end near the Hannaford store is a great place to check. The habitat around the Portland Breakwater Lighthouse (aka "Bug Light") often lures songbirds during migration, especially early in the morning. A walk along Eastern Promenade, which overlooks the ocean, can be very productive even in winter, when a variety of winter ducks and gulls may be present.

BRADBURY MOUNTAIN STATE PARK AND OTHER HAWK-WATCHING HOT SPOTS
In recent years this site near Freeport has become Maine's best-known and only staffed spring hawk-watching location. From March to May, a short hike will lead you to a lookout that, when winds are from the south or southwest, can be a prime spot to watch lots of northward migrating hawks pass by. Great fall hawk-watching locations include Mount Agamenticus in York, Mount Megunticook in Camden, and Beech Mountain on Mount Desert Island.

POPHAM BEACH STATE PARK AND REID STATE PARK

These are two of Maine's most popular state parks. Famous for their beaches, both parks are located just south of the historic shipbuilding city of Bath, on the ends of twin peninsulas on opposite sides of the Kennebec River. Popham Beach State Park, next to neighboring Morse Mountain Preserve, is particularly good during migration. Look for gulls here out over the water. Reid State Park has a wonderful mix of rocky headlands, beach habitat, saltmarsh, and spruce woods. Warblers and sparrows sometimes abound in the shrubby areas and woodland, and at almost any season there are Common Eiders and Common Loons, as well as gulls offshore and in the lagoon behind the beach.

BOOTHBAY AND DAMARISCOTTA/PEMAQUID REGIONS

Traveling north on Route One, just after you pass the famous Red's Eats and cross the long bridge in Wiscasset, you are at the north end of the Boothbay/Edgecomb peninsula. Making your way south on Route 27 will lead you through Edgecomb and toward Boothbay Harbor. Scattered across the peninsula are more than fifteen preserves and some thirty miles of scenic hiking trails, most courtesy of the Boothbay Region Land Trust and its members. All of the preserves provide good birding, though our favorites include the Ovens Mouth Preserve, Zak Preserve, and Ocean Point Preserve. Look and listen for Black-throated Green and Blackburnian warblers, Barred Owls, Hermit Thrushes, and Golden-crowned Kinglets.

Continuing north a few miles farther on Route 1 will bring you to the postcard-perfect village of Damariscotta, which sits at the top of the Pemaquid Peninsula. The Damariscotta River Association has a number of preserves that offer great hiking trails in the area, including the Damariscotta River Farm, where the organization's headquarters are located. The hayfields and ponds below the farm host ducks and a variety of songbirds. Pemaquid Point, at the southern tip of the peninsula, offers stunning open ocean vistas and views of Monhegan Island, and is a fabulous spot to watch for coastal birds. During the summer, daily boat trips to see Atlantic Puffins leave from nearby New Harbor.

MONHEGAN ISLAND

Located ten miles offshore from Pemaquid Point, Monhegan Island is famous among birders as a spring and fall migration hot spot. During April and May in the spring and August through October in the fall, the island can have thousands of birds passing through or stopping off for food and rest. An abundance of these birds can be easily observed at close range, making it a great place for a beginner—yet because there are always a scattering of rarities, the island is a favorite among experienced birders as well, many of whom are happy to share their expertise. During the busy seasons, you can take one of several boats over in the morning and return in the late afternoon, but an overnight stay makes for a less hurried and more memorable visit, especially since every day on Monhegan can reveal different species that arrived during their night migration.

AUGUSTA AREA

The region around Maine's state capital is blessed with many great birding locations. One of the most popular is the Viles Arboretum on the east side of the Kennebec River, where during summer it is easy to see Tree Swallows and Eastern Bluebirds bringing food to the young in their nest boxes. Chestnut-sided Warblers, Baltimore Orioles, Rose-breasted Grosbeaks, and many other species are common and easy to see here. Behind the Augusta State Airport, the City of Augusta maintains a natural area that can be accessed from the back of the cemetery that adjoins the airport. In late May and June, look for Indigo Buntings, Yellow Warblers, and Field Sparrows.

Just north of Augusta, the south end of Messalonskee Lake is home to Pied-billed Grebes, Common Loons, ducks, and other species, and it is home to one of the state's few breeding colonies of Black Terns. The Kennebec Land Trust offers more than forty properties in the greater Augusta area that are terrific for birds. These include the Mt. Pisgah Conservation Area in Winthrop, Vaughan Woods in Hallowell, and the Parker Pond Headland Preserve in Fayette.

RANGELEY LAKES REGION

A few hours' drive inland from the coast brings you to the Rangeley Lakes region, an area famous since the late 1800s as a destination for boating, fishing, hunting, and enjoying the clean air and water of Western Maine. With the surge in the popularity of birding, we can now add the sport of birding to the list of reasons why people visit Rangeley. The area boasts an incredibly beautiful landscape of spruce/fir and hardwood forests, crystal clear lakes, and smooth dark rivers and streams. There are many places to look for birds, including hiking trails of varying lengths and difficulties. Many birders come to the region to look specifically for so-called boreal specialty species like Spruce Grouse, Boreal Chickadee, Gray Jay, and Black-backed Woodpecker. One of the best places to search for these is on Boy Scout Road off of Route 16 in Oquossoc. But where there are uncommon species, familiar birds also are present, and you can find "northern" warblers here, such as Blackburnian, Black-throated Blue, Magnolia, and Nashville. Dark-eyed Juncos and White-throated Sparrows breed in this area, and the region hosts an abundance of thrushes, including Veery and Hermit Thrush.

BAXTER STATE PARK

This 209,000-acre park in north-central Maine, within which lies Maine's tallest peak, Katahdin, was an incredibly generous gift from former Maine governor Percival Baxter. Not surprisingly, a park this size is home to an amazing number and variety of birds, including the boreal specialty species mentioned above. There are many miles of roads and trails throughout the park, including the well-known trails to reach the summit of Katahdin, and wonderful camping areas. During spring and summer, listen for the *I'm so la-zee* song of the Back-throated Blue Warbler, abundant here in the beech forests, and the *Quick-three-beers* song of the Olive-sided Flycatcher ringing out from the top of a dead tree in a boggy swamp.

ORONO BOG/BANGOR CITY FOREST

With its 4,200-foot-long boardwalk through a large sphagnum bog and nearly ten miles of hiking and biking trails, the Orono Bog/Bangor City Forest has become very popular for birders and recreationists of all sorts. The site is only a few miles off the Bangor Mall exit of Interstate-95, so it is easily accessed and a great spot to stop at when passing through. Palm Warblers, Dark-eyed Juncos, White-throated Sparrows and other sparrows, as well as a range of additional species nest in the bog.

MOUNT DESERT ISLAND—ACADIA NATIONAL PARK

The 47,000-acre Acadia National Park, with its islands, mountains, rocky shores, lakes, ponds, and more than 100 miles of paths and trails, is the gem of this spectacular region. There are abundant places to look for birds within the park and in other areas on Mount Desert Island. Places like Pretty Marsh and Ship Harbor are great for nesting songbirds in summer, including Blackburnian Warbler, Black-throated Green Warbler, Purple Finch, Golden-crowned Kinglet, and many others. Sieur de Monts Spring is a fabulous spot to check during spring migration. From July through September, the Bar Island Bar (a sandbar exposed at low tide) right in Bar Harbor often hosts a variety of sandpipers, plovers, gulls, and other waterbirds.

SCHOODIC POINT

Just to the east of Mount Desert Island, Schoodic Point juts out into the ocean like a long finger. Much of the peninsula is within a portion of Acadia National Park, and its many trails and turnouts are also great places to look for birds.

DOWNEAST REGION/LUBEC, EASTPORT

Bird enthusiasts from other parts of Maine and throughout the U.S. come to the eastern-most parts of coastal Maine because the region's unspoiled lands and waters support an impressive diversity of birds and other wildlife. It is not unusual for birders to see thousands of sandpipers and plovers of more than a dozen species on the Lubec Flats from July through September. Tens of thousands of gulls and terns flock to the waters in the area, especially at Campobello Passage in late summer and fall. Seabirds like Black-legged Kittiwakes and Northern Gannets can often be seen along with Humpback, Finback, and Minke whales off West Quoddy Head in Lubec and East Quoddy Head on Campobello Island (you will need a passport to travel from Lubec, Maine, to Campobello, which is in New Brunswick). A great variety of warblers and boreal-forest-dwelling specialties like Boreal Chickadee and Spruce Grouse occur in the area in places like Quoddy Head State Park, Moosehorn Wildlife Refuge, and the Cutler Coast Public Reserve Lands. Be prepared to see a wide range of warblers in this region, including Palm, Magnolia, Black-throated Green, Nashville, and many others.

MORE BIRDING HOT SPOTS
You can find an extensive list of birding sites in Maine in several books dedicated to the subject, including *A Birder's Guide to Maine* by Liz and Jan Pierson and Peter Vickery, and the *Maine Birding Trail* by Bob Duchesne, which is also online at www.mainebirdingtrail.com.

TAKE IT FURTHER
There are ample opportunities across Maine to join birding trips through Audubon chapters, bird clubs, land trusts, and birding stores. Maine Audubon has an extensive lineup of field trips for all levels of interest. Many state parks, Acadia National Park, and national wildlife refuges offer bird trips and programs. A number of commercial bird guiding options are now also available throughout the state, and there are birding festivals in a number of areas. More extensive week-long sessions are offered at Audubon's Hog Island camp in Bremen, Maine, and through the Humboldt Field Research Institute at Eagle Hill in Steuben, Maine. Search online for information about any of these exciting opportunities.

TAKE ACTION FOR BIRDS

Most people who are interested in birds also want to know what they can do to help protect them. From everyday actions to participation in annual ornithological projects, there are many ways people of all ages and skill levels can get involved.

SPEAK UP FOR GOOD POLICIES The voice legislators want to hear most is their constituent's. Call, write, or e-mail your elected officials when laws and policies are up for vote that will have an impact on birds or their habitats. Urge them to vote for protections. To stay informed of such policies, join the Natural Resources Council of Maine, the state's most effective nonprofit organization working for policies that protect Maine's environment.

SUPPORT BIRD EDUCATION EFFORTS Maine Audubon offers many opportunities for hands-on learning about birds. From activities at their Gilsland Farm headquarters in Falmouth to Hog Island camp sessions where many of North America's top birders offer week-long workshops, taking advantage of any of these opportunities is a great way to show you care about birds.

JOIN A LAND TRUST Land trusts and other land-conservation groups provide the invaluable service of preserving land that includes a wide range of habitats needed by birds for feeding, breeding, and raising their young. There are dozens of land trusts doing great work throughout Maine. Support those in your area or on whose land you visit regularly.

VOTE FOR LAND FOR MAINE'S FUTURE This program provides funds ensuring protection of land in all sixteen Maine counties, including habitat for birds and recreation areas for

enjoying them. This matching-grant program stretches funds as far as possible. Urge your legislators to support land bonds, and then vote for them when they make it onto the ballot.

LIVE AN ENVIRONMENTALLY AWARE LIFE Drive a fuel-efficient car (and drive less), make your home energy efficient, buy less stuff, shop for locally grown foods, support local businesses, and look for other ways to tread lightly on the planet.

BECOME A CITIZEN SCIENTIST You don't have to be an ornithologist to contribute data that helps birds. "Citizen science" combines your bird observations with those of others so that researchers can determine changes in bird populations over time and geographic areas. Citizen science projects are excellent ways to get kids interested in birds and provide a fun, educational activity for the whole family. Citizen science projects include:

- **CHRISTMAS BIRD COUNT** Join tens of thousands of volunteers across Maine and the Americas who count birds over specific geographic ranges for one day between December 14 through January 5 as part of this annual bird census. This longest-running citizen science project is run by the National Audubon Society.

- **GREAT BACKYARD BIRD COUNT** A project of the National Audubon Society and the Cornell Lab of Ornithology, the "GBBC" invites birders to spend at least fifteen minutes watching birds during President's Day weekend each February. Count the highest number of each bird species you see at any one time and report your tallies at www.birdcount.org. There you can also view maps to see how your sightings fit in with others pouring in from around the continent.

- **EBIRD (WWW.EBIRD.ORG)** Keep track of the birds you see year-round and explore sightings from around the world online. Your observations will be used by bird watchers, scientists, and conservationists to learn more about the distributions and movements of birds.

- **VISIT THE CORNELL LAB** of Ornithology's website to learn about more projects you and your family can participate in. You'll also find out about online ornithology courses you can take and much more at www.birds.cornell.edu.

Just as important as all of these activities is sharing your passion. Take a child out birding. Visit a classroom, day care, or retirement home and talk about birds or give a presentation. Let restaurants and coffee shops know when you are spending your money in their establishment because you are in the area to bird. The more people there are who care about birds, the more momentum there is for efforts to protect them and their habitats.

ABOUT THE AUTHORS

Jeffrey V. Wells and Allison Childs Wells are native Mainers whose families in Maine go back hundreds of years. Both lifelong birders, they began birding together when they met in college. After graduating from the University of Maine at Farmington, they went on to graduate programs at Cornell University in Ithaca, New York, where Jeff received his M.S. and Ph.D. and Allison, her M.F.A. Both stayed on at Cornell, Allison as communications director for the world-renowned Cornell Lab of Ornithology, and Jeff, also at the Lab, as director of Bird Conservation for National Audubon (first for New York State then for the U.S.). They returned to Maine in 2004 to raise their child among family and Maine's spectacular natural environment. They have published hundreds of bird-related articles and have collaborated on many projects, including as co-contributors to the *Sibley Guide to Bird Life and Behavior* and as creators and webmasters of the websites Arubabirds.com and Bonairebirds.com, which provide field identification and bird-finding information about the birds of these popular vacation islands. Jeff is also author of *Birder's Conservation Handbook: 100 North American Birds at Risk,* published in October 2007 by Princeton University Press, and editor of *Boreal Birds of North America,* published in 2011 by University of California Press. Allison was coeditor of *Birder's Life List and Diary* (third edition) and contributed many bird family accounts to Scholastic's *New Book of Knowledge.* Jeff is now senior scientist for the International Boreal Conservation Campaign and is a visiting fellow at the Cornell Lab of Ornithology. Allison is senior director of public affairs for the Natural Resources Council of Maine. They live in Gardiner with their son and two bird-watching indoor cats.

ABOUT THE ILLUSTRATOR

Evan Barbour earned his bachelor's degree from Reed College with a thesis on animal play behavior and a credential in scientific illustration from UC Santa Cruz. Following a stint at the Cornell Lab of Ornithology, his illustrations have adorned magazines such as *National Parks Magazine* and books such as *South Pacific Birds.* Also an educator with tenure at the California Academy of Sciences, Evan's current pursuit of a MFA at Mills College further enables him to explore the intersection between art and science. Wherever his pursuits take him, Evan maintains an avid interest in birding. To view more of his work, visit www.evanbarbour.com

INDEX